THE MOON

A celebration of our celestial neighbour

THE MOON

Published by Collins
An imprint of HarperCollins Publishers
Westerhill Road
Bishopbriggs
Glasgow G64 2QT
www.harpercollins.co.uk

In association with Royal Museums Greenwich, the group name
for the National Maritime Museum, Royal Observatory,
the Queen's House and *Cutty Sark*
www.rmg.co.uk

Image credits see page 254

A catalogue record for this book is available from the British Library

ISBN 978-0-00-828246-2

10 9 8 7 6 5 4 3 2 1

Typeset by Gordon MacGilp
Printed by RR Donnelley APS, China

If you would like to comment on any aspect of this book,
please contact us at the above address or online.

e-mail: collins.reference@harpercollins.co.uk

facebook.com/CollinsAstronomy

@CollinsAstro

MIX

Paper from responsible sources

FSC™ C007454

This book is produced from independently certified
FSC™ paper to ensure responsible forest management.

For more information visit: www.harpercollins.co.uk/green

GREENWICH

THE MOON

A celebration of our celestial neighbour

Edited by
Melanie Vandenbrouck
Megan Barford
Louise Devoy
Richard Dunn

THE MOON

Contents

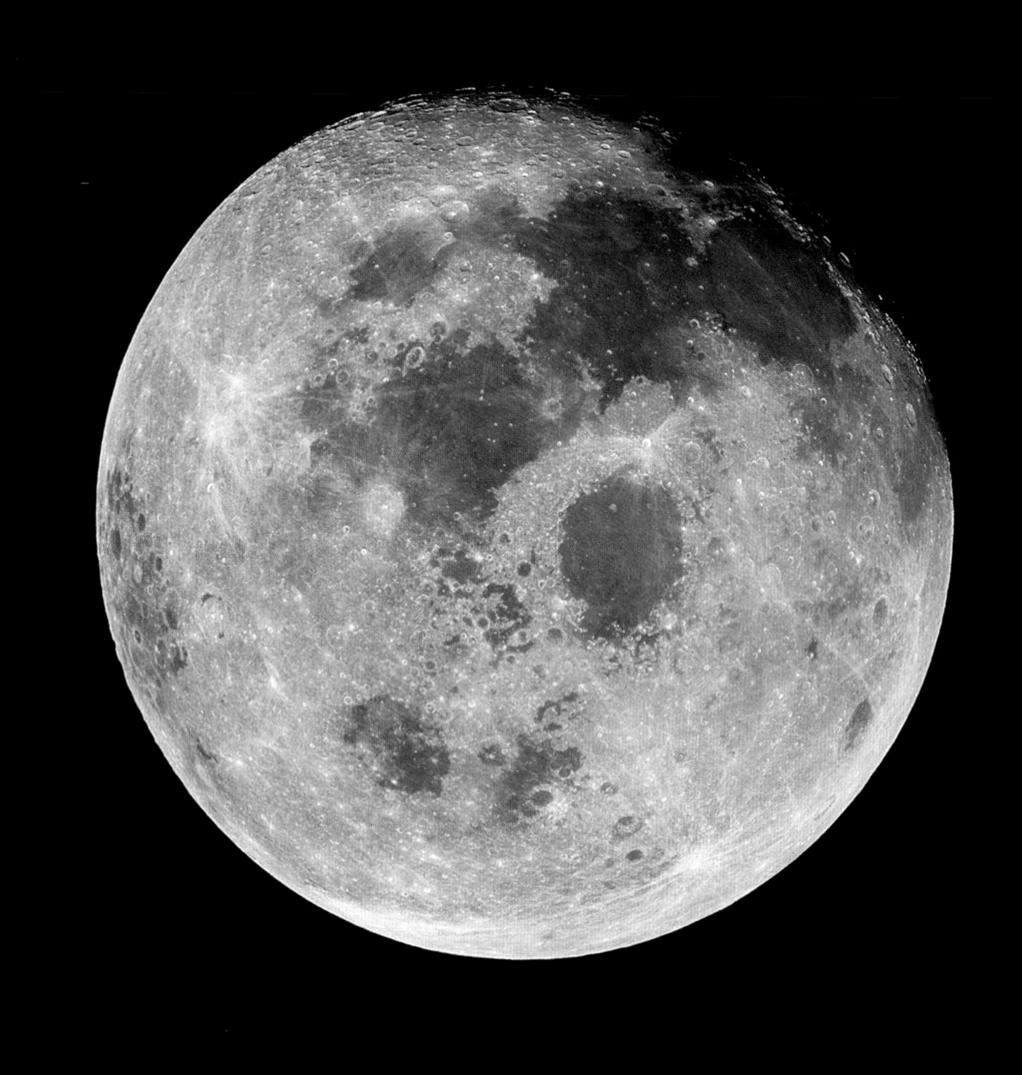

Director's Foreword

Kevin Fewster, AM, Director

Our close relationship with the Moon spans the whole of human history, across all times and cultures. For thousands of years it was a distant light in the sky, remote and unattainable in all but the wildest dreams. In July 1969, it became a tangible place when humankind first stepped onto its dusty surface.

Greenwich has long been linked to the Moon. As a maritime hub governed by the rising and falling of the tides, Tudor kings built their Palace of Placentia here for its position on the Thames, and in 1675 the Royal Observatory was founded by Charles II to track the lunar motions in order to improve the navigation of ships. Two centuries later, our Royal Observatory was a pioneering centre for astrophotography, with the Moon as a prominent target to capture on photographic plates. Today, the Royal Observatory along with the Queen's House, the National Maritime Museum and *Cutty Sark*, form Royal Museums Greenwich. As a family of museums whose sites and collections speak to the collisions and conjunctions of science, art and trade, in which the drive to explore and chart territories looms large, Royal Museums Greenwich is an obvious place to commemorate the 50th anniversary of the Apollo 11 Moon landing.

As we celebrate this momentous feat of technology and collective enterprise that extended the frontiers of space, it is also a time to reflect on how we have projected our dreams and desires onto the Moon. As our constant companion, it has exerted enduring influence, practical as well as philosophical and religious. As the closest celestial object to Earth, the Moon has been painstakingly observed, drawn and mapped. While it long remained untouched by human activities, it became a site on which satire and imagination could run free, before becoming the theatre for Cold War rivalries. It is the world beyond the Earth we know most intimately, yet it still harbours no end of secrets. Even in the twenty-first century, our complex relationship with the Moon continues to evolve.

The Moon is shared by us all and has seen the whole of human history unfold under its light. This book and the exhibition it accompanies are an opportunity to rethink this relationship afresh.

● *Full Moon photographed from the* Apollo 11 *spacecraft on its journey home to Earth*

Introduction

Is the Moon a ball of air and fire, its dark patches a result of their rippling interaction? Do we see blotches on the Moon because of imperfections in our eyesight? Does the Moon act as a mirror, reflecting Earth's image? Or is the Moon a solid orb, a world beyond the Earth?

These were the questions which animated the dialogue of Greek scholar Plutarch's (45–126 CE) *De facie orbis lunae*, 'On the Face of the Moon'. In a lively debate, Plutarch's five scholars reasoned that the Moon wasn't made of air and fire: the colours during a lunar eclipse suggested a body that was solid and Earth-like, not one that could emit its own light. Its appearance couldn't be due to imperfect sight either: those with better eyesight saw the blotches more sharply. A mirror? No: the dark patches were broken, not continuous, and bore little resemblance to the continuous ocean then believed to surround the known world. Anyway, the Moon's face looked the same from places far apart. Discussing the Moon's light, the shadows the Earth cast on it, and which it cast on the Earth, they concluded that the Moon must, in fact, be a solid body. Determined to understand the purpose for which the Moon had been made, they concluded that it was a visible, physical Hades, the resting place of departed souls.

Today, we know that the Moon is a solid sphere that orbits the Earth while the Earth orbits the Sun. As the chapters in this book will show, in some ways each of the questions in Plutarch's dialogue still animate how we think about the Moon. The dark patches on the Moon are not a result of flawed vision, but as astronomers, artists, storytellers and writers have reflected, what you see *does* depend on how you look. The Moon is not a ball of fire and gas. But what it is made of, how it was formed, and what that might tell us about the origins of the Solar System, are questions that continue to motivate scientists today. The Moon is not a mirror of the Earth but it is a mirror for human dreams, obsessions and endeavours. We look to the Moon and find ways to mark time, to create order in a disordered world. We look to the Moon and use it to tell stories about love and loss. We look to the Moon and shape our ambitions, wondering if we could get there and, if we did, what we might do.

Today we are less likely to think of the Moon as an object created for a particular reason or purpose, and even less likely as the final dwelling place of souls. Indeed, in the early twenty-first century, we are perhaps all too aware of the destruction that can be wrought by asking too insistently what use something might have *for us*. Because what of the Moon's future? Will it be one of commercial exploitation and resource extraction? Will the Moon be a site for international competition, as it was during the Space Race? Could it be a space for cooperation and collaboration instead?

In this book, we explore the myriad ways in which the Moon, our nearest celestial neighbour, is bound up with life on Earth. Presenting a scientific and cultural history of our satellite, the different authors offer perspectives from art, science, exploration, discovery and mythology.

The first section explores how the Moon has always been deeply embedded in human culture. People all over the world have used it to locate themselves in time and space, developing lunar calendars and modes of navigation that made use of Earth's satellite. As a constant companion, bright and close, it has been woven into myth and served as a source of artistic and philosophical investigation and inspiration.

The second set of essays reflects on how, following the invention of the telescope in 1608, our view of the Moon was changed forever, becoming a place with its own landscape and features. The resulting observations led to astonishingly beautiful and technically detailed depictions, which in turn allowed the human imagination to take flight. With the advent of photography in the mid-nineteenth century, the Moon's features came into ever sharper view, although this way of looking brought new challenges alongside new opportunities.

The third group covers fictional and factual lunar travel. Since classical antiquity, the Moon has regularly played a role in fiction, often providing a safe stage for biting satires of earthly society. From the nineteenth century onwards, attention was turned to imagining, and then devising, realistic ways of travelling there, beginning with the ground-breaking science fiction of Jules Verne and H.G. Wells. The endeavour to escape Earth's gravity was absorbed into fashion, design, film and art. It culminated in the Apollo 11 Moon landing on 20 July 1969.

The fourth and final section considers our complex and ever-changing relationship with the Moon in the later twentieth and twenty-first centuries. The Space Race involved much sacrifice and, even before the last astronauts left the lunar surface on Apollo 17 in 1972, the value of landing on the Moon was being questioned in different forums: poets and politicians, artists and activists (who were sometimes the same people) wondered what would come out of so costly an endeavour. Today, we know more and more about the lunar body, its geological makeup and its influences on terrestrial life, but it feels like the future of the Moon has never been more uncertain.

Because the Moon is so present to people all around the world, the content of this book is necessarily selective. We acknowledge that the Moon has a role in most belief systems, in most societies, whilst recognising that we cannot do justice to all of them. What we can do, however, is suggest that certain fundamental questions – concerning the ways in which people look to the Moon to understand their place in the Universe, and how different societies have bound the Moon into their ways of life – have been asked, and are still being asked, all over the world.

We can also ask about the implications of our answers. As the ambitions of governments and private enterprise re-focus on the Moon, understanding how people across the world respond to our satellite will only become more important.

A CONSTANT COMPANION

The Moon has always been intimately linked with human culture. We have observed it and made sense of its changing shape for millennia, from simply watching its phases to experiencing the awe and wonder of temporary darkness during a solar eclipse.

Associated with deities and legends worldwide, the Moon has become a part of all human life, both as a spiritual presence and a useful celestial companion. Scholars studied the Moon's place in the cosmos, and began to recognise that the lunar motions could be used to mark time, aid navigation, predict the tides and perhaps even the future. Similarly, generations of artists, writers, poets and songwriters have been inspired by the Moon to explore human emotions and experiences.

Today, technology has enhanced our understanding of our celestial neighbour, but has also reduced its significance. Cocooned in artificially-lit homes and governed by atomic time, we may feel that our relationship with the Moon is now remote, but perhaps these enduring artistic responses may help us reconnect with the elusive and capricious nature of the Moon and its light.

Louise Devoy and Richard Dunn

◯ Gazing at the Moon: Observations and Observatories

Louise Devoy

The Moon is one of the brightest objects in the sky and easy to observe with a simple glance upwards. Yet, over the centuries, for a variety of purposes, people in different civilisations have built observatories and instruments to study its changing appearance and positions, generally alongside a range of other functions. This chapter explores four types of observatory. We begin over 5,000 years ago, with the seasonal and lunar alignments at Stonehenge, then move eastwards to the medieval Islamic world, where state rulers invested much effort and funds into the construction of observatories to help define their calendar based on lunar months. Returning to England in the seventeenth century, we will review the origins of the Royal Observatory, Greenwich, as astronomers endeavoured to improve navigation at sea. Finally, our story will focus on the mid-twentieth century use of radio telescopes at Jodrell Bank in the UK to track the first missions to the Moon.

Symbolic significance

Without written records it is difficult to assess how our early ancestors observed the Moon, but certain clues still remain in the landscape. One of the most famous examples is Stonehenge, a group of large stones situated on the Wiltshire plains of southern England. Although the monument was most likely built as a place of ritual, burial and social gatherings, the distinctive alignments in the site's layout suggest a link between the stones and observations of the Sun and Moon. Starting about 5,000 years ago, the creators of Stonehenge somehow recorded the changing position of these celestial bodies over many decades, perhaps initially setting wooden stakes in the ground before adding stone markers at a later date. For example, once a year on the summer solstice around 21 June, the Sun appears to rise above the so-called 'Heel' stone that lies along a ridge, known as the Avenue, which extends towards the northeastern horizon.

For the Moon, there are two patterns of change. Firstly, over the course of a lunar month (29·5 days), the position of moonrise moves along the eastern horizon from northeast to southeast. Secondly, as the Neolithic observers at Stonehenge realised, the extent of this monthly changing position also varies, but over a cycle of about 18·6 years. Consequently, four 'station stones' within the circular ditch were aligned with the most extreme positions of moonrise and moonset during this longer cycle. Although we have little insight into the motivation of our Neolithic ancestors, we can still marvel at their ability to connect this place of ritual to the changeable motion of the Moon.

A natural calendar

Gradually changing in appearance over the course of 29·5 days, the Moon provides a natural calendar, and its repeating pattern has been exploited across many cultures. In the Islamic world, the official adoption of the lunar calendar in the years following the death of the Prophet Muhammad in 632 CE continued the traditional nomadic use of the Moon for navigation and timekeeping in the desert. The Islamic lunar calendar consists of 12 months of 29 or 30 days, which means that it runs approximately 11 days behind the solar year, hence the dates of festivals successively move through the civil Gregorian calendar. The most important observation is the sighting of the Crescent Moon, just two days after New Moon, which denotes the start of the new month. Many larger mosques had their own astronomer, known as the *muwaqqit*, who was responsible for observing the Moon to keep track of the calendar and for calculating the times of the five daily prayers according to the Sun.

This approach was perfectly adequate for day-to-day timekeeping, but not for more ambitious rulers. Inspired to improve the accuracy of the astronomical data tables known as *zijes*, successive generations of rulers invested in elaborate observatories with large instruments at Maragha (1259), Samarkand (1424), Istanbul (1577) and Jaipur (1738). These were centres of scholarship with international teams of astronomers and facilities such as libraries and schools. Apart from the well-preserved instruments of Jaipur, little of these sites remains visible today. Nonetheless, colourful illustrations of these once impressive observatories

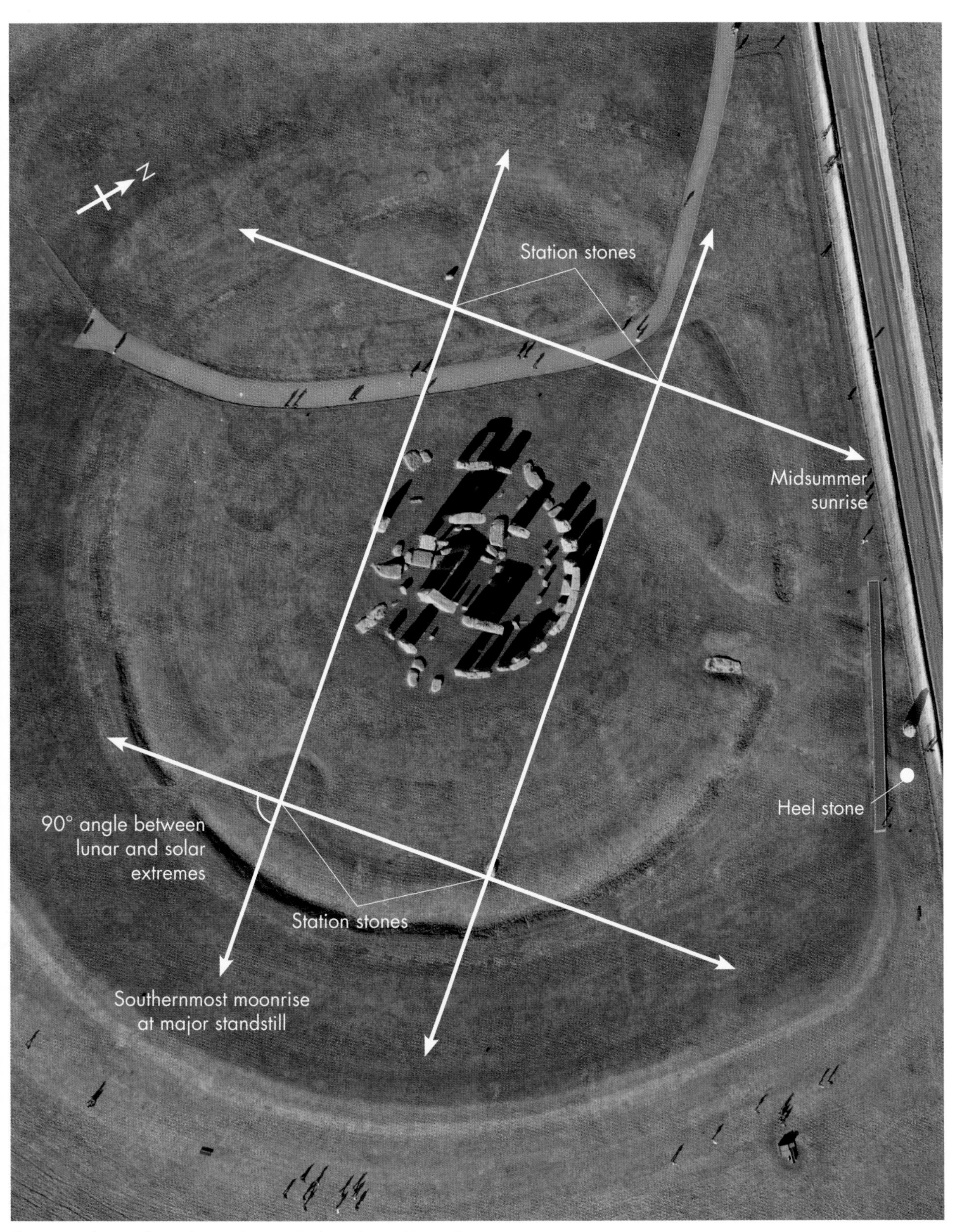

An aerial view of Stonehenge showing the key solar and lunar alignments

Illustrated manuscript showing astronomers at work in the observatory in Istanbul in 1581
Artist unknown, Shahinshahnama ('Book of the King of Kings') by Ala al-Din al-Mansur, Istanbul University Library, MS.F.1404, folio 57a

Huge stone instruments used for measuring the position of the Sun, Moon and stars at Jantar Mantar Observatory, Jaipur, India, 2008

demonstrate the importance of the Moon and astronomy across the Islamic world.

Sea and sky

Lunar observations have also been a powerful navigational tool, above all through the development of the lunar-distance method in which mariners measured the angular distance between the Moon and certain stars to assess their local time and position. By the mid-seventeenth century, with global trade routes well established and European nations vying for control of the seas, trying to make the lunar distance method a viable technique was becoming a political and economic priority. In England, Charles II appointed an 'astronomical observator' in a royal warrant dated 4 March 1675 'for perfecting the art of Navigation'. Three months later, the king signed a second warrant, ordering the construction of an observatory on royal land in Greenwich.

John Flamsteed (1646–1719), a young man from Derby, was chosen as the first Astronomer Royal and began his lunar observing even before the new observatory was ready. From April to July 1675 he was at the Tower of London, using the northeast turret of the White Tower to house two long telescopes (24 feet (7·3 metres) and 8 feet (2·4 metres)), pendulum clocks and other instruments. In the autumn he moved to Greenwich and observed from the Queen's House, the Palladian mansion designed by Inigo Jones in 1616. With the new observatory completed in the summer of 1676, Flamsteed continued to compile his star catalogue, the *Historia Coelestis Britannica*, which was eventually published posthumously in 1725. Unfortunately, the unwieldly format of the catalogue and the complex calculations needed to use it made it impractical for most navigators. Later, they came to rely on the *Nautical Almanac*, first published by the fifth Astronomer Royal, Nevil Maskelyne, for the year 1767. With new instruments for measuring the required angles in the sky, and improved timekeepers that functioned well at sea, the *Nautical Almanac* made the lunar distance method feasible and cemented the observatory's role within global navigation.

Racing towards the Moon

From the mid-twentieth century, lunar observation began to focus on helping to plan and track the first tentative journeys to the Moon itself. One observatory involved was Jodrell

The Royal Observatory from Crooms Hill, *about 1696*
Artist unknown, English school

Bank in Cheshire, the brainchild of physicist Bernard Lovell and engineer Charles Husband, who used their wartime radar research experience to convert scrap battleship parts into radio telescopes. The largest was the enormous 76·2-metre telescope (now known as the Lovell telescope) that was completed just a few months before the launch of *Sputnik 1* by the USSR in October 1957. Although the telescopes were intended to study cosmic rays, meteors and radio sources in deep space, attention shifted towards the Moon as Russia and America competed to demonstrate their supremacy during the Cold War.

Situated midway between the two rivals, the astronomers at Jodrell Bank were requested by both space agencies to track lunar probes. In February 1966, the Jodrell team was able to intercept signals from the Russian probe *Luna 9*, which had successfully landed on the Moon and was transmitting back to Earth. Realising that the radio signals were in fax machine format, the team borrowed a picture receiver from the *Daily Express* newspaper and the first images from the lunar surface appeared as headline news the next morning.

Scientists at Jodrell Bank also witnessed key moments of the Apollo programme during July 1969. Fluctuations in the signals collected by the 50-foot radio telescope clearly showed the moment when astronaut Neil Armstrong took manual control of the Apollo 11 *Eagle* lander to steer the craft towards a safe landing site. Meanwhile, the Lovell telescope was being used to track the unmanned Russian probe *Luna 15*, whose mission to bring back lunar soil samples before Apollo 11 was thwarted when it crashed into the lunar surface. Although these tracking missions were just

Journalists cluster around the fax machine at Jodrell Bank Observatory to receive images of the Moon from Luna 9*, 7 February 1966*

a small part of the research at Jodrell Bank, the observatory was a first-hand witness to the triumphs and failures of the Space Race.

Our motivations for observing the Moon have changed from using it as a calendar in the sky, to developing navigation, to supporting the exploration of it. Looking to the future, the Moon may itself become a giant observatory as astronomers based on its far side use its body to mask unwanted radio interference from Earth and listen further into space. Still, the achievements of each culture are remarkable and remind us of how much the changing rhythm of the skies has been entwined with people's lives. From Stonehenge to the Apollo programme, humans continue to be inspired to observe the Moon and develop new technologies to enhance our use and understanding of our closest neighbour in the Solar System.

○ Using the Moon

Richard Dunn

People found uses for the Moon well before it came within reach. Its constant presence meant that almost every civilisation used it in the governance of daily life. Its movements and phases have helped mark time and navigate the oceans. Its influence on those oceans has led to speculation and, eventually, understanding of what causes the tides. More controversially, people everywhere have tried to understand whether the Moon affects their actions, perhaps even their futures. Some explored whether understanding astrological influences might allow them to be used for the benefit of humanity.

Marking time

> *It was He that gave the Sun his brightness and the Moon her light, ordaining her phases that you may learn to compute the seasons and the years.*
>
> Qur'an 10:5

The Moon's changing appearance and movements are so obvious and seemingly regular that it can be no surprise that every civilisation has thought to use it for calendrical purposes, often in conjunction with the solar cycles of days and seasons. The earliest attempts to mark time from the Moon may date back to the Neolithic period, though there is some debate about the evidence. Nevertheless, what soon became clear to anyone observing for a reasonable time is that the Moon goes through a full set of phases from new to full and back again every 29·5 days or so. This cycle is the origin of the months used in almost every calendar. For the same reason, the words 'moon' and 'month' share a common root in the English and Germanic languages.

Yet, despite their seeming regularity, the Moon's motions are remarkably complex and do not synchronise in a simple way with the apparent motions of the Sun. Trying to resolve the differences between lunar and solar time has guided centuries of close study and mathematical modelling to create working calendars, whether for agricultural, religious or civil purposes. By the beginning of the first millennium BCE, Babylonian astronomers had developed the observing skills and mathematics to model the lunar and solar movements and produce calendars in which each year began on the first New Moon after the spring equinox. Their observations divided the Moon's changing appearance into quarters, called *shabbattu*, which became the Sabbath of the Jewish calendar and the basis of our seven-day week. While other civilisations developed similar systems, the Babylonians went further, studying lunar motions over very long periods to identify repeating patterns in order to make predictions about the future.

The challenge for any calendar was to make the cycle of lunar months somehow match the Sun's so that the two did not move out of phase. The problem is that the Moon's cycle is not a whole number of solar days, which led to often piecemeal solutions. By the fifth century BCE, for instance, the Babylonians had created a lunisolar calendar based on a cycle of 19 years, each year comprising 12 lunar months with extra months inserted seven times in each cycle to equate to 19 solar years. Similar calendars were devised by other cultures, including China and ancient Greece, with those still in use including Hebrew, Jain, Buddhist and Hindu religious calendars. In the Hebrew calendar, for example, the first day of the month, Rosh Chodesh, is marked by the New Moon and is a minor religious festival. Other cultures abandoned the Moon and moved to purely solar calendars, as the Egyptians did in the third millennium BCE.

The Roman Empire made a similar change in 46 BCE under Julius Caesar, aided by the otherwise unknown astronomer Sosigenes of Alexandria. Caesar's calendar replaced a lunar one based on a year of 12 months totalling 355 days. The older calendar had therefore needed many intercalations – days or months added to keep pace with the Sun's movements – that were dictated by priests. However, these were not applied systematically, thus creating major problems in what came to be called the 'years of confusion' just before the reform. The new Julian calendar, in which a year averaged 365·25 days, achieved with a leap year every fourth year, was meant to solve the problem permanently and keep the years in step with the Sun without the need for further intercalations. However, its introduction did mean that 46 BCE had to be a massive 478 days to bring everything back in line.

TRANSPARENT DIAGRAM OF

THE PHASES OF THE MOON.

The various appearances which the Moon periodically presents in her revolution round the Earth, are termed *Phases*, and arise from the different positions which its opaque mass assumes in relation to the Sun and the Earth. When the Moon is between the Sun and the Earth, its dark side is presented to us, and it is consequently invisible; in this position it is called the *New Moon*. Four days after the time of New Moon, it has receded 45 degrees from the Sun, and now a portion of its illumined surface is seen in the form of a crescent (fig. 2.) After eight days, it has departed 90 degrees from the Sun, and shows a bright semi-circular disk; the Moon is now said to be in its first quarter. Gradually shewing more of its illumined surface, it becomes gibbous (fig. 4); and about fifteen days after the time of New Moon, it stands

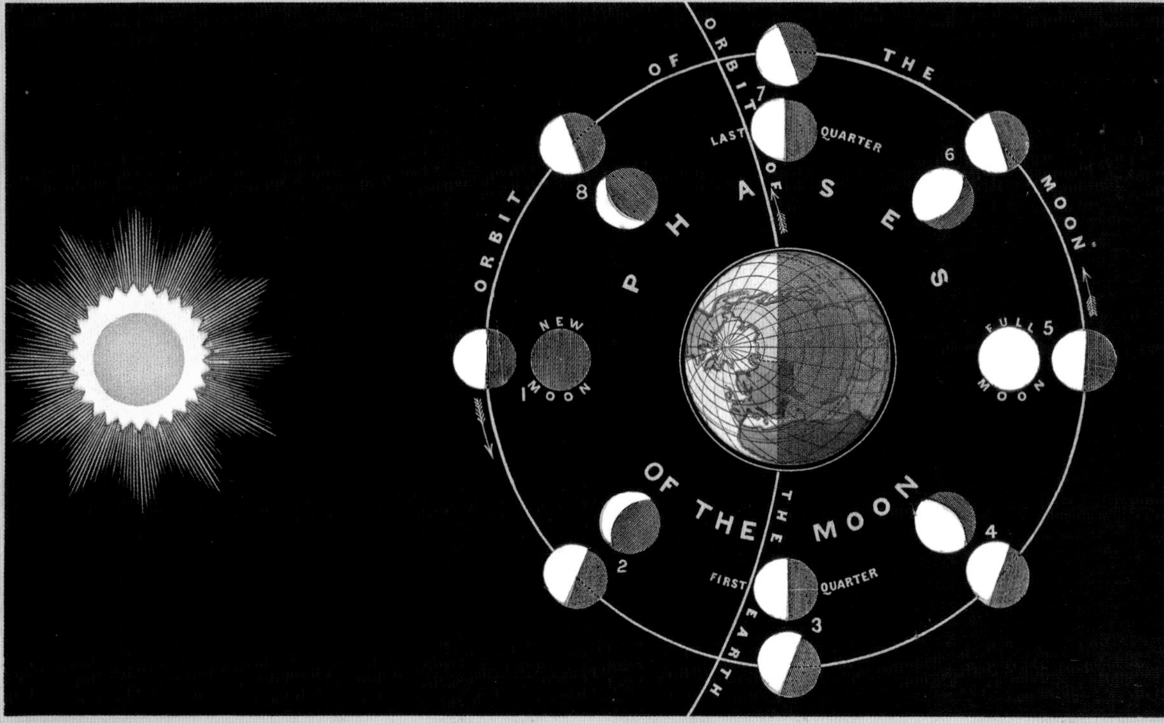

directly opposite the Sun, presenting a complete circular disk; this is the *Full Moon*, rising when the Sun sets, and shining through the whole night. Proceeding in its course, its illumined surface gradually decreases; approaching the Sun, it becomes a second time gibbous (fig. 6); a Half Moon at its third quarter, assumes a crescent from (fig. 8); and completing its orbit, disappears, becoming a New Moon again as at first. The pointed ends of the Moon's figure when a crescent are called its Cusps or Horns. During the first quarter, they point to the eastward, or the direction in which the Moon is moving in its orbit; and in the last quarter, to the westward; or that path which it has just described.

The *apparent* motion of the Moon is that of rising in the east, and setting in the west; but this is owing to the revolution of the Earth upon its axis. The Moon's real motion round the Earth is from west by south to east. It moves at the rate of forty miles per minute. The Moon turns once on its axis every month, and therefore, always presents the same side towards the Earth; from one-half of the Moon, therefore, our world is always visible; whilst, from the opposite hemisphere, it can never be seen. The Earth, as seen from the Moon, appears thirteen times as large as that body appears to us, and reflects thirteen times the quantity of light that the Moon reflects to us. Experiments have shown, that the light of the full Moon is three hundred thousand times less than that of the Sun; and that it produces no heat, for if its rays, concentrated by a powerful mirror, be thrown on the bulb of a thermometer, no effect is perceptible. The mean distance of the Moon from the Earth is about two hundred and thirty seven thousand miles, and its diameter is computed to be 2160 miles.

LONDON: PUBLISHED BY JAMES REYNOLDS, 174, STRAND.

'Transparent Diagram of the Phases of the Moon' by James Reynolds, c.1850

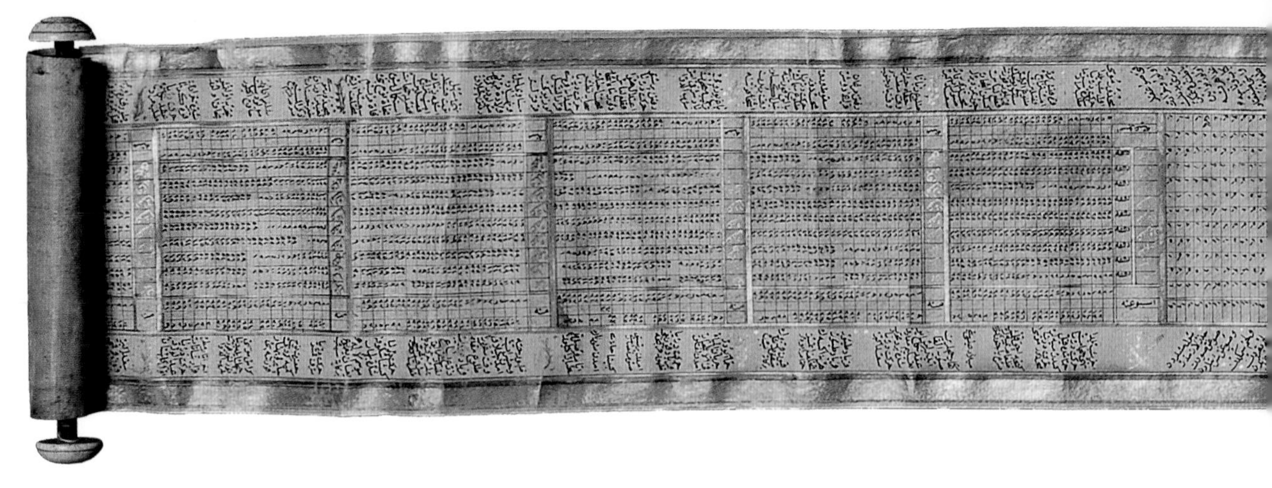

Islamic lunar calendar on parchment, Turkey, 1795–96

By contrast, the Islamic calendar has remained purely lunar. It began to be adopted after the death of the Prophet Muhammad in 632 CE and became a unifying force as the Islamic world expanded. As with the Julian calendar, one of the aims was to abandon the practice of intercalation, which was needed with the lunisolar calendar previously used in Arabia. Taking the *hijra* (Year of Migration, 622 CE in the Western calendar) as its start, the calendar has 12 lunar months of alternately 29 and 30 days, totalling 354 or 355 days, with each month beginning about two days after the New Moon. The Moon's central place in the calendar, which required the New Moon to be actually sighted to mark the beginning of each month (a practice that continues today for the religious calendar), stemmed from the words of the Qur'an and became a spur for important work in astronomy and mathematics. By the ninth century CE, astronomers based largely in Baghdād had defined a 30-year cycle of 12 lunar months that matched the true lunar orbit to within a day in 2,500 years.

While Christian territories used a solar calendar, the Moon was crucial for Christianity's defining festival, Easter. Its dating is linked to Jewish Passover, as Christ's crucifixion and resurrection were meant to have happened then. This tied Easter to the lunar cycle, meaning that the date drifted within the solar year. With different Christian factions failing to agree on a dating method, the Roman emperor Constantine (272–337 CE) used the Council of Nicaea of 325 CE to push through changes that tied the dating to the spring or vernal equinox and thus to the Sun.

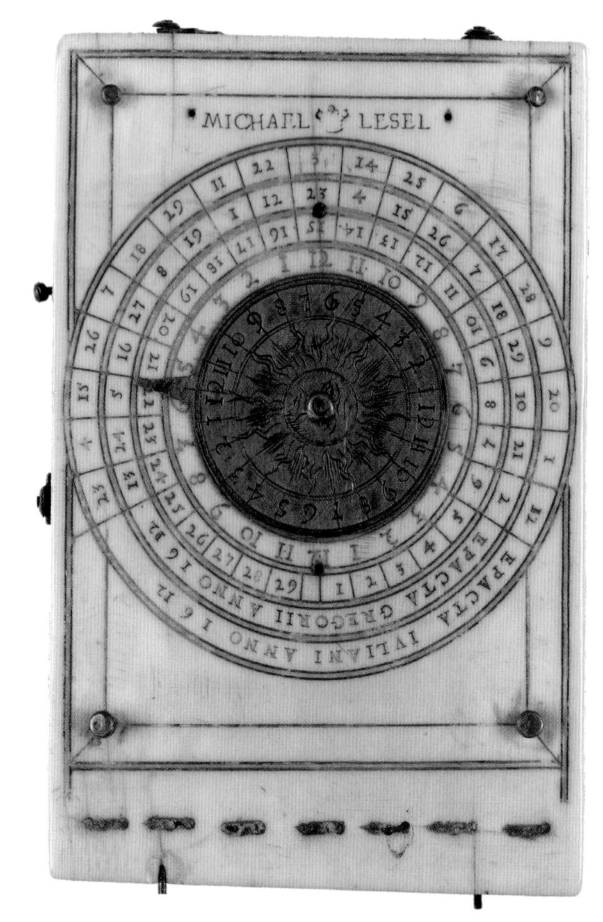

Diptych dial, by Michael Lesel, c.1612, with scales for calculating the date of Easter according to the Julian or Gregorian calendars

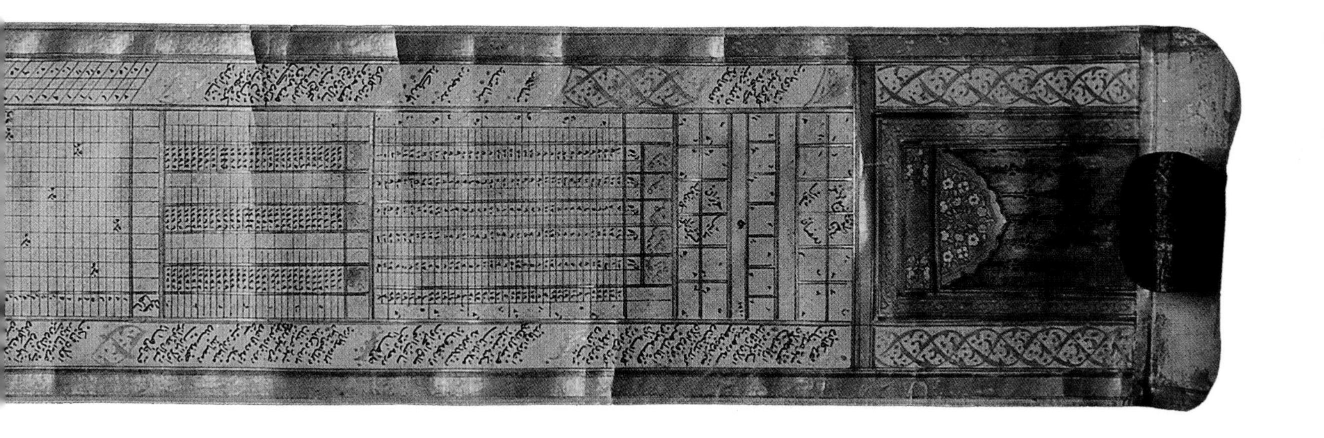

Thereafter, Easter was to be on the first Sunday after the first Full Moon on or after the equinox. Nonetheless, the dating of Easter remained a key motive for observing and modelling the Moon and its movements. Indeed, the early Christian theologian and philosopher, Saint Augustine (354–430 CE) wrote that this was astronomy's only useful purpose. Later theologians, including the English monk Venerable Bede (672/3–735 CE), continued devoting great effort to the calendar and Easter's date. Yet, as most churches had fixed the vernal equinox at 21 March, its date continued to drift within the solar year. By 1267, the English philosopher Roger Bacon (1214–92) complained that the true equinox was on 12 March and that holy days such as Easter were now being miscalculated. The debates would continue, with the Moon and Easter driving calendrical reform right through to the introduction of the Gregorian calendar from 1582, which Britain adopted in 1752. Other countries, including Greece and Russia, only adopted the calendar in the twentieth century, thus making it important to be able to work with both calendars for over 300 years.

Space and place

The Moon has long allowed people to move around at night as a source of illumination, particularly around Full Moon. Indeed, this gave the name to the Lunar Society of Birmingham, a learned society that met for half a century from the mid-1760s. The Society, whose members included engineer James Watt, entrepreneur Josiah Wedgwood and natural philosopher Joseph Priestley, gathered each month on a day close to the date of the Full Moon, when the extra light would make the journey home easier and safer in an age without street lighting.

In navigation, the Moon has had a more complex role, particularly at sea, due to its gravitational effect on the Earth and its use in timekeeping. Today we understand that the Moon's gravitational pull is the main cause of the tides. As early as the fourth century BCE, the Greek geographer and explorer Pytheas noted that the height of the twice-daily tides seemed to depend on the lunar phase. Nevertheless, coming up with a physical explanation for these observations took hundreds of years, as different thinkers proposed a range of mechanisms. Living as a monk not far from the river Tyne, Bede had ample opportunity to observe the changing tides and thought that the Moon was a cause. He also noticed that each port had its own unique tidal characteristics. Five hundred years later, the Persian philosopher Abu Yahya Zakariya' ibn Muhammad al-Qazwini (1203–85) proposed that solar and lunar heating caused the tides. The German mathematician and astronomer Johannes Kepler (1571–1630) thought it was a magnetic pull, while the French philosopher René Descartes (1596–1650) invoked a lunar vortex. Others including Galileo Galilei (1564–1642) offered non-lunar explanations. It was only with the publication of Isaac Newton's (1642–1727) theories that a comprehensive physical mechanism was put forward, one that recognised the importance of the Moon's gravitational pull and with which French mathematicians including Pierre de Laplace (1749–1827) were able to formulate a convincing analysis.

A page from a nautical almanac on vellum, by Guillaume Brouscon, c.1546, showing the establishments, indicated by lines from the compass rose, of French ports including Calais

A table of the establishments of various ports (right) on a calculating device known as Napier's bones, c. 1679

Even without a physical mechanism and mathematical model, it was possible to produce tables and instruments that sailors could use to predict the tides, thus allowing ships to enter or leave port at the right times to avoid shallow waters and adverse currents. The predictions were based on the phase and direction of the Moon. First, the sailor needed the 'establishment' of the port in question, a concept put forward by Bede. This was the compass bearing of the Full or New Moon at high tide for that port. By knowing the number of days since Full or New Moon, sailors could calculate when high and low tide would take place, and were usually aided by lists of establishments and associated tide tables, which appeared in books of sailing information. Simple instruments were also developed to perform the calculations.

Other, more sophisticated ways of using the Moon made it into a celestial timekeeper designed to help determine longitude (east-west position), which, when combined with the latitude (north-south position) found from observing the Sun or a star, allowed you to pinpoint your location on a map. Because longitude is measured in the direction of the Earth's rotation, the longitude difference between two places can be thought of as the difference between their local times, i.e. the time given by the Sun's position in the sky at that place, local noon occurring when it is at its highest that day. As the Earth rotates through 360° in 24 hours, one hour of time difference between two places is equivalent to 15° of longitude. Alternatively, one could say that Earth turns through one degree of longitude every four minutes. Many longitude methods therefore aimed to compare the local time, from the Sun or another star, at the observer's location and, at the same time, at a reference point of known geographical position. Knowing the local time at a place far away from the ship was tricky, as it meant either carrying that time using a portable timekeeper – a technical challenge largely solved by John Harrison in the mid-eighteenth century – or using the movements of the Moon or another celestial object as a 'celestial clock' to find it.

The Moon can be used to find time – and thus longitude – in several ways. One is from eclipses. If an eclipse has been predicted to occur at a specific local time for a known location, someone observing it from another place knows the time at the reference location and can compare it to their own local time for the eclipse. This was well known in Ancient Greece and there were attempts to use it for position-finding and mapping by the late-fifteenth century. The explorer Christopher Columbus observed two eclipses in the Caribbean in 1494 and 1503, hoping to establish more accurate longitudes, but his results were poor. Others including the Welsh explorer Thomas James (in 1631) and Englishman John Wood (in 1670) made similar attempts on later voyages of exploration. Far more ambitious was a project of the 1570s and 1580s that sought to map Spain's territories in South America by having local officials build simple moon dials, mark where the Moon's shadow fell at the beginning and end of each eclipse and send the results back to Spain. Sadly, the returns proved too difficult to process and were full of errors. Philipp Eckebrecht's world map of 1630 was more successful in its use of lunar eclipse data to fix the longitudes of key places. It was the first map to equate one hour of time, astronomically determined, to 15° of longitude on Earth.

Eclipses are too infrequent for day-to-day navigation, but other lunar observations did have potential. The most fully developed method was that of lunar distances or 'lunars', which was described as early as 1514 by German mathematician Johannes Werner (1468–1522). The principle was to treat the Moon's rapid motion relative to the stars like the hand of a clock. The critical measurement was the angular distance between the Moon and the Sun or a star, the so-called lunar distance. Having also measured the altitude of the two bodies (their angular height above the horizon), the navigator

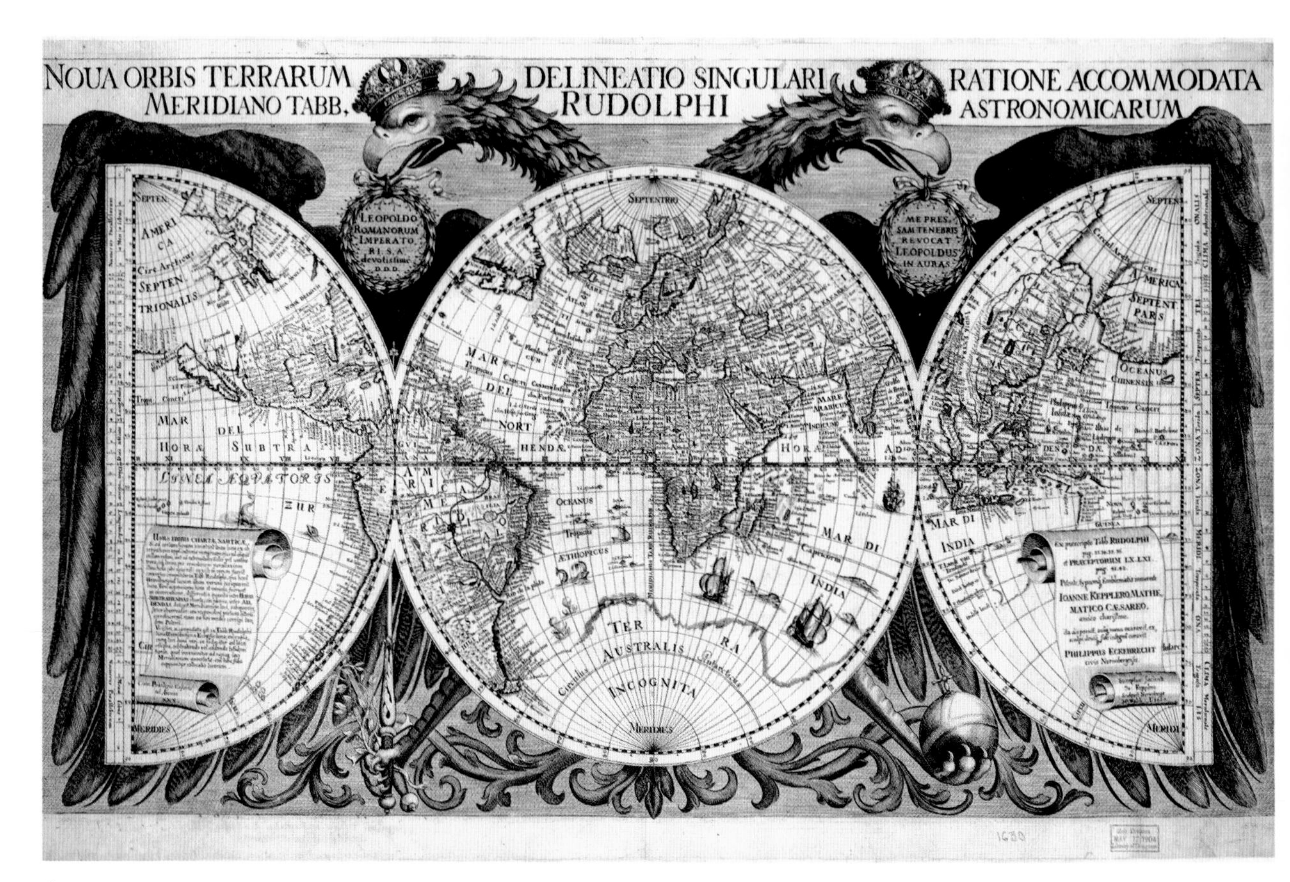

World map, by Philipp Eckebrecht, Nuremberg, 1630, drawn using lunar eclipse data from the Tabulae Rudolphinae, a set of astronomical tables by the astronomer Johannes Kepler, first published in 1627

performed a series of calculations and, with the aid of tables of the Moon's future positions, worked out the time for the reference point. They could then calculate the longitude of their vessel by comparison with their own local time. In theory, it was straightforward. What took time and effort was developing tables that could predict the Moon's positions several years ahead and instruments able to make accurate measurements from a pitching and rolling ship.

To the casual observer, the Moon's motion may appear regular but the gravitational interaction of the Sun, Moon and Earth – the basis of what is known as the three-body problem – make it remarkably complex. State support for work towards longitude methods also included the promise of large rewards, beginning in Spain in 1567 and later including the British Longitude Act of 1714, which would finally lead to the development of practicable longitude solutions, including a method for lunars. This was no easy matter. Creating a mathematical model of the Moon's motions had defeated the brightest European mathematicians for generations. Isaac Newton's *Principia Mathematica* (1687) laid the foundations for a powerful mathematical analysis of celestial motions, yet even he failed to crack the three-body problem. As he confessed, 'his head never ached but with his studies on the moon'.[1]

It was Newton's followers who finally tied the Moon down. The breakthrough came from a Hanoverian astronomer, Tobias Mayer (1723–62). Mayer began with the work of the leading mathematicians of his time, including Leonhard Euler, Jean d'Alembert and Alexis Clairaut, who had been applying new forms of analysis to the three-body problem. Crucially, Mayer combined this with a detailed study of the best observations, including the Royal Observatory's. Bringing theory and observation together allowed him to devise an

○ *Observing the Moon with a sextant, from John Lawrence King, log of the Owen Glendower, 1846–47*

○ *Measuring the lunar distance, from Petrus Apianus,* Introductio Geographia *(Ingolstadt, 1533), title page (detail)*

[82] JULY 1769.

Diſtances of ☽'s Center from Stars, and from ⊙ weſt of her.

Days.	Stars Names.	Noon. ° ′ ″	3 Hours. ° ′ ″	6 Hours. ° ′ ″	9 Hours. ° ′ ″
1	α Arietis.	36. 9. 45	38. 0. 20	39. 51. 7	41. 42. 7
2		50. 59. 8			
6	The Sun.	38. 37. 36	40. 14. 12	41. 50. 22	43. 26. 7
7		51. 18. 32	52. 51. 44	54. 24. 32	55. 56. 56
8		63. 32. 56	65. 2. 56	66. 32. 32	68. 1. 46
9		75. 22. 46	76. 49. 57	78. 16. 48	79. 43. 21
10		86. 51. 44	88. 16. 35	89. 41. 11	91. 5. 33
11		98. 4. 19	99. 27. 28	100. 50. 27	102. 13. 16
12		109. 5. 21	110. 27. 24	111. 49. 21	113. 11. 13
13		119. 59. 37			
10	Regulus.	48. 49. 12	50. 20. 34	51. 51. 46	53. 22. 46
11		60. 54. 57	62. 24. 50	63. 54. 34	65. 24. 9
12		72. 50. 13			
12	Spica ♍	18. 48. 44	20. 17. 10	21. 45. 36	23. 14. 2
13		30. 36. 15	32. 4. 40	33. 33. 6	35. 1. 33
14		42. 24. 7	43. 52. 42	45. 21. 21	46. 50. 2
15		54. 14. 37	55. 43. 47	57. 13. 3	58. 42. 24
16		66. 10. 43	67. 40. 44	69. 10. 53	70. 41. 10
17		78. 14. 38			
17	Antares.	32. 46. 4	34. 16. 56	35. 47. 59	37. 19. 12
18		44. 57. 56	46. 30. 14	48. 2. 42	49. 35. 22
19		57. 21. 30	58. 55. 17	60. 29. 15	62. 3. 24
20		69. 57. 4			
20	β Capricorni.	15. 9. 39	16. 44. 42	18. 20. 0	19. 55. 32
21		27. 56. 37	29. 33. 27	31. 10. 29	32. 47. 44
22		40. 57. 13			
22	α Aquilæ.	47. 35. 59	48. 55. 24	50. 15. 49	51. 37. 11
23		58. 36. 17	60. 2. 15	61. 28. 48	62. 55. 57
24		70. 19. 36	71. 49. 43	73. 20. 13	74. 51. 6
25	α Pegaſi.	34. 43. 29	36. 17. 24	37. 52. 14	39. 27. 55
26		47. 36. 52	49. 16. 23	50. 56. 23	52. 36. 51
27		61. 5. 26	62. 48. 12	64. 31. 16	66. 14. 38
28		74. 55. 29	76. 40. 20	78. 25. 23	80. 10. 36
29	α Arietis.	45. 39. 33	47. 27. 43	49. 16. 2	51. 4. 27
30		60. 7. 51	61. 56. 40	63. 45. 29	65. 34. 18
31		74. 37. 51			

JULY 1769. [83]

Diſtances of ☽'s Center from Stars, and from ⊙ weſt of her.

Days.	Stars Names.	12 Hours. ° ′ ″	15 Hours. ° ′ ″	18 Hours. ° ′ ″	21 Hours. ° ′ ″
1	α Arietis.	43. 33. 18	45. 24. 38	47. 16. 4	49. 7. 34
6	The Sun.	45. 1. 27	46. 36. 21	48. 10. 50	49. 44. 54
7		57. 28. 55	59. 0. 31	60. 31. 43	62. 2. 31
8		69. 30. 39	70. 59. 12	72. 27. 24	73. 55. 15
9		81. 9. 35	82. 35. 32	84. 1. 13	85. 26. 37
10		92. 29. 42	93. 53. 39	95. 17. 24	96. 40. 58
11		103. 35. 56	104. 58. 28	106. 20. 53	107. 43. 11
12		114. 33. 0	115. 54. 44	117. 16. 24	118. 38. 1
10	Regulus.	54. 53. 34	56. 24. 11	57. 54. 37	59. 24. 52
11		66. 53. 36	68. 22. 55	69. 52. 7	71. 21. 13
12	Spica ♍	24. 42. 28	26. 10. 55	27. 39. 21	29. 7. 48
13		36. 30. 0	37. 58. 29	39. 27. 0	40. 55. 33
14		48. 18. 47	49. 47. 37	51. 16. 32	52. 45. 32
15		60. 11. 51	61. 41. 24	63. 11. 3	64. 40. 50
16		72. 11. 35	73. 42. 9	75. 12. 51	76. 43. 41
17	Antares.	38. 50. 35	40. 22. 9	41. 53. 54	43. 25. 50
18		51. 8. 13	52. 41. 16	54. 14. 29	55. 47. 54
19		63. 37. 45	65. 12. 17	66. 47. 1	68. 21. 57
20	β Capricorni.	21. 31. 18	23. 7. 18	24. 43. 31	26. 19. 58
21		34. 25. 12	36. 2. 52	37. 40. 46	39. 18. 53
22	α Aquilæ.	52. 59. 25	54. 22. 29	55. 46. 21	57. 10. 58
23		64. 23. 41	65. 51. 57	67. 20. 41	68. 49. 55
24		76. 22. 20			
24	α Pegaſi.	28. 39. 30	30. 8. 28	31. 38. 53	33. 10. 36
25		41. 4. 24	42. 41. 36	44. 19. 26	45. 57. 52
26		54. 17. 46	55. 59. 7	57. 40. 52	59. 22. 59
27		67. 58. 17	69. 42. 13	71. 26. 24	73. 10. 50
28		81. 55. 59			
28	α Arietis.	38. 28. 41	40. 16. 5	42. 3. 43	43. 51. 33
29		52. 53. 0	54. 41. 38	56. 30. 19	58. 19. 3
30		67. 23. 7	69. 11. 53	71. 0. 36	72. 49. 16

○ *Pages from the* Nautical Almanac *for 1769, showing the time at Greenwich relating to positions of the Moon for July 1769*

improved theory and new lunar and solar tables. Their accuracy would earn him a posthumous reward under the Longitude Act.

Although he had never seen the sea, Mayer worked out how to improve shipboard observation as well, devising a circular observing instrument that was tested at sea in the 1750s and 1760s. The trials would lead to the development of the marine sextant, a lynchpin of navigation for the next two hundred years. They also led to the creation of the *Nautical Almanac*, which followed a model proposed by the French astronomer Nicolas Louis de Lacaille (1713–62) for annually published tables of the Moon's future positions. First published for 1767, each volume was calculated from observations at the Royal Observatory and became a model for publications in other countries.

Yet, even with the new instruments and tables, the lunar distance method was arduous. It could be reduced to a series of steps to be mechanically followed, looking up the data and corrections in printed tables that simplified the spherical geometry, but the method was most suited to mathematically-minded sailors. James Cook learned lunars from astronomer Charles Green on his first Pacific voyage (1768–71) and became a firm advocate. Others found it tiresome. Speaking of a 1795 naval expedition, the former governor of New South Wales, Thomas Brisbane, recalled 'not ten individuals who could make a lunar observation'.[2] Yet, with the development and gradual adoption of longitude-finding by chronometer, lunars were able to complement other forms of navigation and were crucial for checking the accuracy of the timekeepers. Only in 1907 did the Nautical Almanac Office stop publishing tables for the lunar distance method, by then redundant following the adoption of new celestial navigational techniques during the nineteenth century and of radio time signalling at the beginning of the twentieth century.

Nevertheless, the Moon kept a place in navigation, as it is one of the objects one can observe to plot navigational position lines and to find latitude from a meridian passage – observing its altitude as it crosses the observer's meridian or north-south line. The Moon is particularly valuable because it can be visible during the day, allowing solar and lunar observations to be used to find longitude. However, because the Moon is close to Earth and its movement is fast and irregular, this only became possible as lunar data became more accurate during the twentieth century.

Following a virtuous circle, lunar observation has also become important for further improving the accuracy of navigational and other tables, in particular through the observation of occultations. These occur when the Moon passes in front of a star or planet as seen from Earth, making them disappear. Accurately timing an occultation gives a precise measurement of the Moon's position relative to the stars and so can be used to measure deviations from its predicted positions, allowing for the refinement of future predictions and tables. From the 1930s, the British Nautical Almanac Office began calculating and publishing data for forthcoming occultations for observatories and other locations worldwide, as well as co-ordinating the observed results from across the globe. Crucial to their work was an 'Occultation Machine', which mechanically generated approximate predictions of when and where occultations would be visible. Staff could then refine the predictions through more laborious calculations, with the Machine saving time and effort by allowing them to concentrate only on the occultations that would be most useful. Lunar occultations are still used in this way, although the calculations are now swiftly performed by electronic computers.

Prediction and plague

> *Many are of opinion that the Air and time of the Moon is to be considered in several Rural Affairs: As that the Increase is the most fit and best time for the killing of Beasts; and that young Cattle fallen in the Increase are fittest to wean; and that it is the best time to plant Vines, and other Fruit-trees; to graft and to prune lean Trees, and cut Wood; to sow Herbs, and gather Tillage, and cut Meadows.*
>
> John Worlidge[3]

Given the evident effects of the Moon, it can hardly be a surprise that people all over the world connected its changing appearance to earthly activities. Medieval astrologers claimed they could make predictions about the weather, pregnancy and mental health from the dark and light patches on the Moon's surface, while others would scry, reading future events from its reflection in water. English folk traditions still common in the nineteenth century linked its phases with anything from the best time to kill a pig and shear sheep to felling trees or picking apples and mushrooms. Another tradition was that rubbing one's hands under the light of a Full Moon would cure warts.

Such thoughts inevitably raised deeper questions and possibilities. What changes might these celestial influences cause and did they follow rules? Might knowledge of how the Moon and other celestial bodies affect the Earth allow one to make predictions about the future? Might their influences be manipulated or controlled? Exploring these questions has generated myriad approaches to what we might call astrology, which has in turn stoked controversy in religious contexts and come under fierce attack with the rise of modern science. The Moon has been central to these discussions and disputes as one of the bodies that manifestly influences the Earth and as the closest and most changeable of the celestial bodies.

Although we will never know when people first began to observe and speculate about the Moon's influence, there is solid evidence of centuries of astronomical records in the kingdoms of Assyria and Babylonia stretching back to at least the third millennium BCE and written down on clay tablets. Some of these tablets, including a set known as the *Enuma Anu Enlil* ('When Heaven and Earth ...'), list solar and lunar eclipses that were believed to be omens of impending disasters. Although human destiny

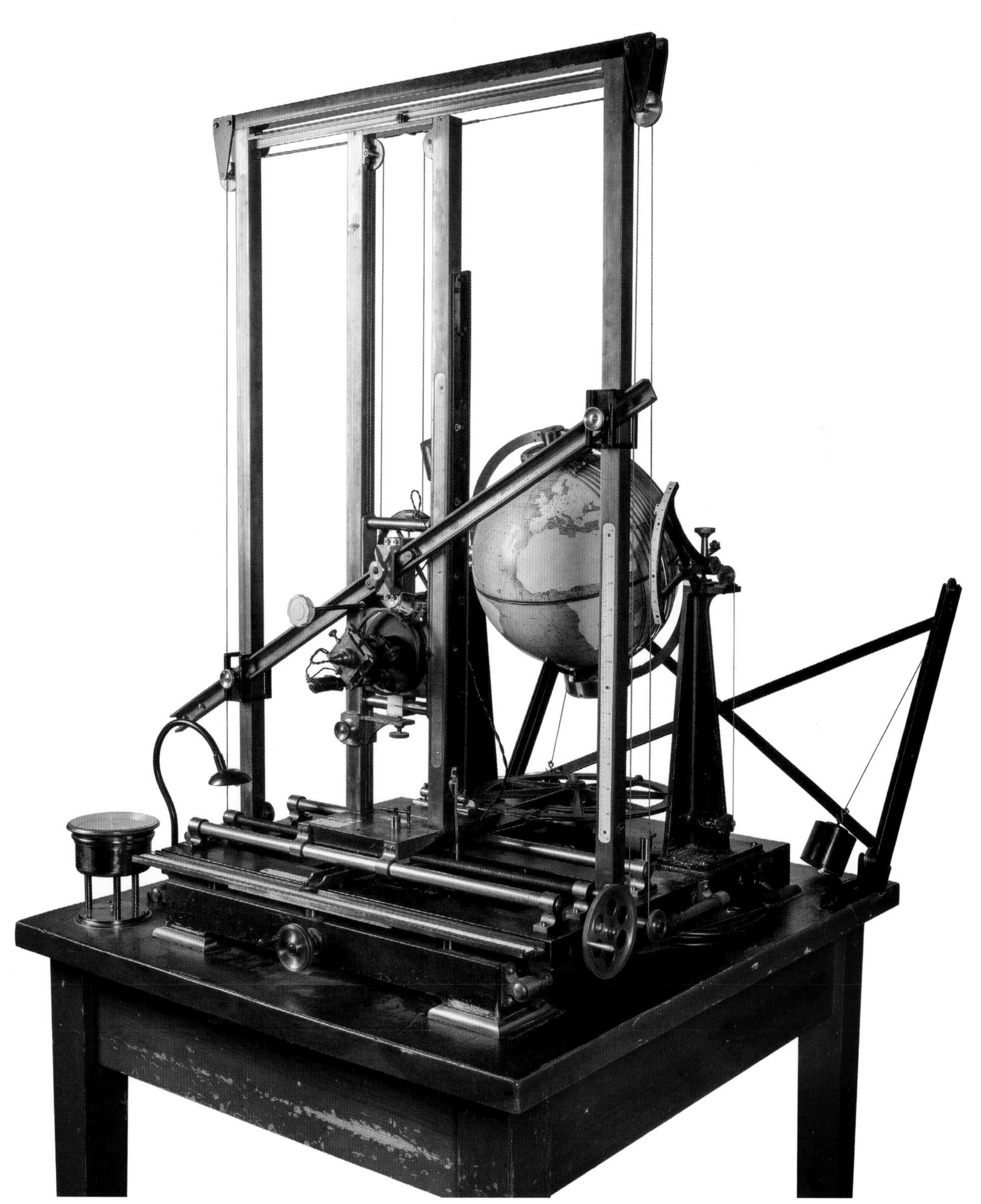

○ *Occultation Machine, made by Arthur Wescott for the Nautical Almanac Office, 1935. It used a car headlamp and a lens to project the image of a star and the Moon's shadow onto a globe marked with key observatories*

was believed to lie in the hands of the gods, these omens could, the Babylonians believed, be used to understand the gods' intentions and perhaps alter them through prayer or other actions.

These early civilisations thus thought of the Moon's and Sun's movements, including eclipses, as signs rather than causes of future events. Other civilisations, by contrast, developed systems in which the stars and planets (with the Sun and Moon considered as planets) were direct causes of terrestrial events. However, there was enormous debate about how predictable and specific these might be. For many, predictions were limited to large-scale phenomena: the weather, agriculture and phenomena affecting large populations, such as disease. In the 1470s, for instance, the Italian scholar and priest, Marsilio Ficino (1433–99), wrote that recent eclipses lay behind widespread outbreaks of plague. Yet while William Shakespeare agreed when he wrote that 'these late eclipses in the Sun and Moon portend no good',[4] the Moon on its own could be a positive influence, as agricultural writers like John Worlidge (1640–1700) suggested.

Restricting astrological prediction to large-scale phenomena avoided a clash with the notion of individual free will, which was so crucial for Christian and other religious doctrine. Nevertheless, as astrology developed from the beliefs of ancient civilisations through the medieval period in Europe and the Islamic world, increasingly sophisticated practices evolved that sought to make predictions about the individual or even to answer specific questions asked at a particular moment. These might cover anything from a person's future, to questions about their health, the success of a particular venture such as a sea voyage, or even crime. In June 1602, Goodwife 'Goody' Green consulted clergyman and astrologer Richard Napier (1559–1634) about a theft. Examining a chart for when Green asked the question, Napier concluded that a servant boy was to blame. In this judicial astrology, the Moon was considered as 'most powerfull in operation of all the other Planets', another seventeenth-century astrologer noted.[5]

It also had a prominent role for Indian, Chinese and Arabic (and so later for European) astrologers, who developed the concept of lunar mansions. This approach divided the Moon's path into 27 or 28 regions, each of which affected the way a planet's influence was exerted when in that mansion.

The Moon had a central role in medicine, one English astrologer writing that a child born at Full Moon would never be healthy, such was its influence. Accordingly, astrology was part of medical training in European universities by the Renaissance. Trainee physicians would learn that the balance of the bodily humours fluctuated with the Moon's phases, with illnesses following a similar pattern. The London consultant astrologer, Simon Forman (1552–1611), noted, for instance, that in general a 'disease will increase as longe as the moone doth goe swifter and when she goeth slower the disease will decrease'.[6] Such ideas underpinned a theory of crises or critical days: a change (crisis) would take place every seven days (i.e. each quarter of the lunar cycle) from the moment the patient took to their bed with the illness. Each crisis marked the patient's improvement or deterioration, while the middle day between crises – the indicative day – might also be a time of significant change. Metal and paper instruments helped the physician to make prognoses according to this system.

As one of the planets, the Moon played a role in other areas of astrological medicine, particularly in helping to decide when best to carry out operations such as bloodletting. The frequently reproduced image of 'Zodiac Man' showed when different parts of the body might best be treated, depending on the Moon's position in the zodiac. Lunar influence might also explain the efficacy of certain medicines and when to gather herbs and administer remedies. Ficino found that 'medicine hardly acts at all when the Moon is in conjunction with Venus'.[7] The renowned English physician and astrologer, Nicolas Culpeper (1616–54), wrote that the plant brank ursine (an acanthus) was good for curing the King's Evil (scrofula) since 'by the influence of the Moon, it reviveth the ends of the Veins which are relaxed'. Celandine, he also noted, was one of the best cures for the eyes when gathered with the Sun in Leo and the Moon in Aries, as 'the Eyes are subject to the *Luminaries* [Sun and Moon]'.[8]

○ *A volvelle used to indicate the course of an illness according to the Moon's age in days, from Petrus Apianus,* Astronomicum Caesareum *(Ingolstadt, 1540)*

47
CANCER
GEMINI
TAVRVS
ARIES
PISCES
AQVARIVS
CAPRICORNVS
SAGITTARIVS
SCORPIVS
LIBRA
VIRGO
LEO
VACVI
INDICATIVA
INTERCIDENS
CRISIS PRIMA
CRISIS SECVNDA
CRISIS TERTIA
PRINCIPIV MORBI
M III

The Dominion of the Moon in Mans Body, passing under the 12 Zodiacal Constellations.

♈ Aries, *Head and Face.*

♉ Taurus, *Neck and Throat.*

♋ Cancer, *Breast, Stomach & Ribs*

♍ Virgo, *Bowels and Belly.*

♏ Scorpio, *Secret members.*

♑ Capricorn, *Knees.*

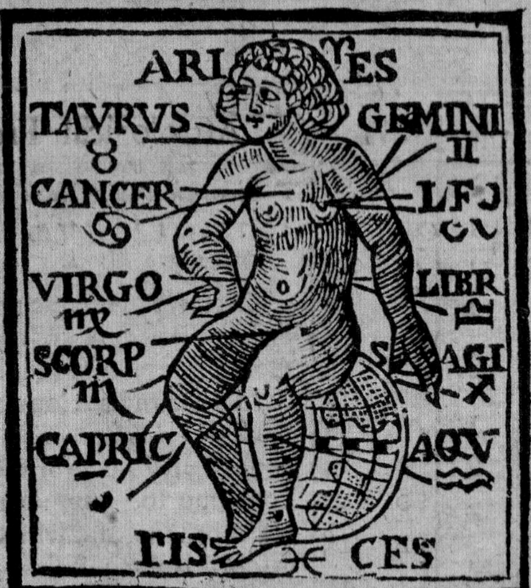

♊ Gemini, *Arms and Shoulders.*

♌ Leo, *Heart and Back.*

♎ Libra, *Reins and Loins.*

♐ Sagitarius, *Thighs.*

♒ Aquarius, *Legs.*

♓ Pisces, *the Feet.*

The Characters of the Seven Planets.

♄ *Saturn*, ♃ *Jupiter*, ♂ *Mars*, ☉ *Sol*, ♀ *Venus*, ☿ *Mercury*, ☽ *Luna*, ☊ *Dragons Head*, ☋ *Dragons Tail.*

Upon the Twelve Signs.

Majestick *Aries* clad in's *Golden Wooll*,
Commands the Head and Face: The Zodiack's *Bull*
Rules Neck and Throat: The Arms and Shoulders lye
Subjected to the Zodiack's *Gemini*:
Cancer the Breast and Stomack do obey,
And Regal *Leo* o're the Heart doth sway
His Royal Scepter, whilst the wanton *Maid*
Claimeth the Belly: Reins and Loins are way'd
By the Cœlestial *Libra*: Secrets are
The poisonous *Scorpions*: And the Thighs the share
Of *Sagitarius*: But the Feeble Knees
Fall to the *Goat*: The Legs are standing Fees
Unto *Aquarius*: Whilst the Feet do bend
Their steps to *Pisces*, and assign an E N D.

These associations reflected complex networks of correspondence that connected everything – from rocks through plants and animals – to the heavenly bodies. To the Renaissance philosopher, the Moon had an affinity with or might act through a multiplicity of things including water, silver, plants such as cabbage, mandrake and poppy, and minerals such as selenite, and was even linked to the angel Gabriel. Astrologically, the Moon was considered matriarchal and ruled over natural cycles related to water, including rain, tides, gestation and childbirth. It governed navigation and other watery trades, such as fishing, as well as the occult arts. As the brain was one of the moistest parts of the body, the Moon was believed to influence it strongly, hence the idea, still discussed in the eighteenth century, that insanity was 'lunacy' or a matter of being 'moonstruck'.

The same correspondences suggested how one might actively exploit and direct lunar influence. According to a lavish compilation of the properties of different gemstones created for King Alfonso X of Castile (1221–84), also known as Alfonso the Wise, selenite follows the waxing and waning of the Moon; a piece hung on a fruit tree encourages rapid growth and ripening. An amulet made of it may also prevent epilepsy. Amulets could capture the lunar influences in other ways. According to the *Picatrix*, a medieval text on magic and astrology that was widely circulated in later centuries, an Indian healer had once cured an Egyptian boy stung by a scorpion by having him drink from a cup containing wax impressed with a seal. The image on the seal had captured the Moon's healing power because it had been inscribed when the Moon was in Scorpio.

Today there are still debates about the extent to which the Moon exerts influence. Some effects are well documented – not only the tides, but also the relationship between coral spawning and lunar phase, and the reduced activity of nocturnal animals around Full Moon. Yet convincing scientific evidence that the Moon affects human biology or mental health has not yet been found. Nevertheless, its more obvious effects on Earth – through light and gravity and due to its regular cycles – have exerted a powerful influence on how humans have led their lives, whether through the ordering of time, the ability to navigate, or attempts to determine what the future might hold. Plans to establish lunar bases now suggest other ways in which we might one day exploit our closest heavenly companion. We will, it seems, always find uses for the Moon.

○ *'The Dominion of the Moon in Mans Body', from Henry Coley,* Nuncius Cœlestis: Or, The Starry Messenger for the Year of Our Redemption, *1681*

◯ 'Never point at the Moon!': Lunar Lore across the Millennia

John J. Johnston

In July 2018, a lunar eclipse or 'Blood Moon' caused huge excitement in the news and social media. Inevitably, heavy cloud in the UK obscured the dramatic spectacle of Earth's shadow sweeping across the Moon's surface, but it presented a timely reminder of the superstitions and folklore associated with the Moon and its many phases and faces. For the Incan empire (1438–1533) of western South America, for example, the Moon's reddening suggested that it was being savaged by jaguars, while in Romania, a form of vampire, the *vârcolaci*, which resemble small dogs or dragons, were thought to spill the Moon's blood as they drank it.

Of all the natural phenomena influencing human culture, the heavens have been a compelling inspiration, with the Moon the most predictably changeable of these. The ways in which humanity has responded to the Moon are endlessly fascinating, particularly as it is most visible in darkness, which in turn affects the human psyche, conjuring not only rest, the comfort of home and sleep, but also monstrous beasts, the spirits of the dead and the fear of one's own mortality. The Moon, in other words, provides a familiar point in an unfamiliar landscape. Consequently, human cultures have negotiated their relationship with the dichotomies of the night through religion, mythology and folk-tales, many of which revolve around the Moon.

One of the earliest extant depictions of lunar worship is the so-called *Venus of Laussel*, (see overleaf) dating from the Upper Palaeolithic period (*c*.25,000 BCE). It is a 44.5-centimetre limestone relief sculpture of a nude female figure, which was discovered in 1911 and originally painted with red ochre; the relief is one of six discovered in the Laussel cavern, a ritual site not far from the famously decorated Lascaux caves in south-western France. In her right hand the figure holds a crescent-shaped object, probably the horn of a bovid, marked with 13 incised notches. The correlation between the shape of the horns and the Crescent Moon is evident in many cultures and historical periods, while the notches possibly refer to the 13 days of the Waxing or Waning Moon or, indeed, the months of the lunar year. Her left hand is placed upon her belly and the whole may, therefore, be related to fertility or to menstruation. The figure may be a lunar deity, the participant in a lunar ritual or a character from an unknown tale.

For clear evidence of Moon-worship or lunar mythology, we must look to written sources. One of the earliest recognisable lunar deities is the Mesopotamian god, Sin/Suen or Nanna. Near-Eastern scholar Mark Hall has argued that the Moon's crescent phases provide the most evident identification of the god, with the Crescent Moon, related to the glowing horns of a bull, being ubiquitous on Mesopotamian cylinder seals. The Crescent Moon perfectly encapsulates the threefold properties of the deity, relating him to the heavens, the procreative strength of the bull and the light of the Moon itself.

Ancient Egypt

Moving westward from the Tigris and Euphrates, ancient Egypt's pantheon of deities includes several gods with lunar attributes or responsibilities. One of the earliest is Iah, whose name means 'Moon'. He is depicted in the *Pyramid Texts* of the Fifth Dynasty (*c*.2392–2282 BCE), where the king declares Iah to be both his brother and his father. However, Iah rapidly becomes associated with the god Khonsu, usually represented as a young man wearing the sidelock of youth and tightly bound in funerary wrappings. He wears a Full Moon resting on an upturned Crescent Moon on his tightly fitting skullcap and holds a crook and flail in his fists. Khonsu was worshipped throughout the Nile valley, although his primary shrine was established in the temple of Amun at Karnak by Ramesses III (1217–1155 BCE). Khonsu's nature altered during the course of Egypt's history from that of a ferocious deity, described in the so-called 'Cannibal Hymn' as assisting the deceased king in capturing, devouring and consuming the strength of other deities, to that of a healer associated with human and animal gestation. It is believed that he was credited with healing Ptolemy IV Philopater (245/4–204 BCE) who went on to adopt the epithet, 'beloved of Khonsu, who protects the king and drives away evil'.[1]

A Tainted Eclipse *by Phil Hart, 2014*

The Venus of Laussel, c. *25,000 BCE*

The most recognisable Egyptian lunar deity is the ibis-headed Thoth, associated with writing, language, wisdom and magic. His lunar characteristic is probably most firmly connected with his responsibility for recording and measuring time. In a fascinating myth, Thoth, 'the reckoner of time and seasons', created the 365-day calendar by beating the inveterate gambler, Iah, at *senet* – an ancient board game similar to backgammon but with religio-mythical associations. Thoth won a portion of moonlight on each occasion, which ultimately totalled five whole days, thereby allowing the sky goddess Nut to give birth to five divine children, Osiris, Isis, Set, Nephthys and Horus the Elder. These epagomenal or additional days made for a more precise calendar and the births of each of the deities became annual festivals. Although Thoth is a lunar deity, he is frequently depicted as a cynocephalus baboon; (see page 38) wily, intelligent animals whose murmuring grunts may have sounded like an unknown language. Principally associated with the Sun, which they greet in the morning with shrieks and upraised hands at sunrise, when associated with Thoth, they are usually portrayed wearing the double Crescent and Full Moon on their heads.

Bronze figurine of Iah with inlaid silver eyes, Late Period

Faience amulet of Thoth as a baboon with Full and Crescent Moon headdress, Third Intermediate Period

○ *Bronze figurine of Osiris with Full and Crescent Moons atop a heavy tripartite wig, Late Period*

○ Endymion and Selene *by Victor Florence Pollet, c. 1870*

Given the vast – geographical and temporal – sweep of Egyptian religion and its related mythologies, the Egyptians were surprisingly sanguine regarding what we would today describe as the non-linear nature of their narratives. Egyptian religion is complex and syncretistic, cheerfully incorporating many different creation tales and deities. Consequently, in spite of devotion to Iah and Khonsu, Egyptians related the Moon's ever-changing phases to the interminable physical, magical, legal and sexual battles between Horus the Younger and his regicidal uncle Set as they wrestled for control of Egypt. During one of these contests, Set, in the form of a black bull, gouged at his nephew's left eye. While Horus's right eye was that of the Sun, Re, the injured left became the Moon in its different phases, healed by the god Thoth and becoming a potent material reminder of mathematical fractions being incorporated into the hieroglyphic script.

Other Egyptian deities have less obvious connections with the Moon; a number of bronze figurines show Osiris, Lord of the Dead, wearing the Full and Crescent Moon atop a heavy tripartite wig. These statuettes tend to be inscribed as 'Osiris-Iah'. It has been suggested that they may, given Osiris's mortuary connotations, equate to the Waning Moon, while Khonsu with his aforementioned sidelock of youth may equate to the Waxing Moon. A stela dedicated to Osiris from the reign of Rameses IV (died 1149 BCE) described the god as 'the moon in the sky; you rejuvenate yourself according to your desire and become old when you wish'.[2] Similarly, the jackal-headed Anubis, embalming deity and guide of the spirits of the deceased to the afterlife, is surprisingly shown with the lunar disk in a number of royal birth-scenes. It has been suggested that Anubis is present as 'a guarantor of [eventual] rebirth'.[3] The connections between the Moon, darkness, death and rebirth present a fascinatingly complex means of addressing the human condition, elevated through the divine nature of the royal births.

The Graeco-Roman world

For the ancient Greeks, Selene was not only goddess of the Moon but also its personification, hurtling through the heavens upon her divine chariot. According to Hesiod's *Theogony* (*c.*700 BCE), she was the daughter of the second generation of Titans, Hyperion and Theia, and was the sister of Helios, the Sun god, and Eos, goddess of dawn. In his *Argonautica* (*c.*300 BCE), Apollonius of Rhodes tells how Selene fell in love with the beautiful youth Endymion, son of queen Calyce and Zeus, king of the Olympian pantheon, following one of his many dalliances. Selene asked Zeus to grant Endymion eternal youth, sleeping perpetually on the slopes of Mount Latmus, where she could visit him nightly. However, Apollodorus of Athens later suggested that Endymion was a willing victim, when Zeus questioned his son he chose an ageless and deathless eternal sleep. Although Endymion is sometimes described as a prince, more often he is a beautiful shepherd, tending his flocks by Selene's light, by which he first attracted her attention. Intriguingly, a later tradition attributed to Pliny the Elder's *Natural History* (77–79 CE) suggests that Endymion was the first astronomer to record the Moon's celestial movements.

A splendid example of the iconography associated with Selene occurs on a silver dish from the so-called Boscoreale Treasure of silverware, jewellery and coins, discovered in 1895 in the remains of a villa near Pompeii. The dish shows Cleopatra Selene (*c.*40–6 BCE), daughter of Cleopatra VII Philopater (69–30 BCE) and Marc Antony (83–30 BCE), and twin to Alexander Helios. After her marriage to King Juba II of Numidia (52 BCE–23 CE), she helped found the city of Caesarea in Mauretania, modern Algeria. The dish shows her as 'Africa,' wearing an elephant scalp headdress in honour of her illustrious Macedonian predecessor, Alexander the Great (356–323 BCE), and holding aloft the Crescent Moon of the deity after which she was named.

Artemis, the chaste goddess of hunting, archery and mistress of beasts, was the twin of Apollo and daughter of Zeus and Leto. She was closely associated with the Moon and often shown in her lunar guise, wearing a long robe and veil with crescent headdress. In Rome, the Moon was related to Diana – equated with Artemis – and Luna, as well as to Juno, wife of Jupiter. The *kalends* of each month, when the New Moon appeared, were dedicated to this powerful goddess, while on the *nones*, a few days later at First Quarter, Juno was recognised as 'Juno Covella', Juno of the Crescent Moon. For the Romans, Luna was the personification of the Moon, shown traversing the skies in her *biga* chariot, drawn by horned oxen. With Sol, she was represented within the official Imperial cult, symbolising the magnitude of Rome's all-encompassing empire. A complex deity, she was little understood; sometimes merely a cipher, sometimes

○ *Gilded silver dish depicting Cleopatra Selene, c. 1–50 CE*

Detail of gilded silver patera showing Luna traversing the skies in her oxen-drawn biga, c.350–400 CE

little more than an epithet for Juno or Diana and on other occasions represented alongside Hecate and Prosperina within the *Diva Triformis*.

Magic and moonlit rites

Hecate is a particularly fascinating deity. Associated with the Moon, night and witchcraft, the goddess may have been adopted into the Greek pantheon from Anatolia. However, her place within that pantheon was somewhat peripheral, appropriately enough for a deity concerned with transitional spaces – doorways, crossroads, gateways, the communications between night and day and the living and the dead. According to Aristophanes's *Plutus* (408 BCE), meals were proffered to Hecate and the restless dead each lunar month, and food offerings were left to her at crossroads on the first night of the New Moon. Her aforementioned *Diva Triformis* image may relate to the Full, Half and New Moons, although later writers have suggested that the trio corresponded to Selene above, Artemis on Earth, and Persephone, queen of the underworld, below. Hecate's insignia was the Crescent Moon and, according to Diodorus Siculus, her light roused the Byzantines on a miserable, moonless night in 340 BCE when Philip II of Macedon (382–336 BCE) mounted a surprise attack. Her

Marble statue of Hecate Triformis, first century CE

Print of Macbeth, Act IV, Scene I, featuring the Three Witches and Hecate
Robert Thew, based on the work of Sir Joshua Reynolds, c.1786

bright light suddenly blazing from the heavens awoke the dogs from their slumbers and Philip's attack was foiled. The Byzantines subsequently erected a statue to the goddess and adopted her Crescent Moon as their symbol. Indeed, Hecate's influence extended far beyond the ancient world. Although her specifically lunar associations appear to have been largely ignored by William Shakespeare (1564–1616), she not only appears in Macbeth (1606), but also informs the image of the play's 'three weird sisters' through her *Diva Triformis*.

Both Julius Caesar (100–44 BCE) and Pliny the Elder (23–79 CE) note the importance of the Moon in Druidic rites and beliefs, involving sacrifices of white, horned bulls beneath the Moon's light. Caesar goes so far as to suggest that the Gauls are descended from Dis, the Roman god of the dead, thereby having a special relationship with night and the Moon. The historian Tacitus (*c.*56–120 CE) records that the leaders of Germanic tribes would meet at either the New or Full Moon – times they deemed to be particularly propitious. This should come as no surprise, however, given that numerous superstitions continue to proliferate throughout the world regarding the New Moon as the ideal moment to begin new ventures, from planting crops to courtship and even cutting hair and nails. It should also be noted that while it suited these noble Romans to mock the superstitions of northern tribes, such beliefs were not so distant from those of the rustic labourers in the fields of Rome.

A world of characters

The appearance of the Full Moon is culturally mutable.

Across East Asia, the lunar markings traditionally show not a human face but a jade rabbit, named Yutu, wielding a mortar and pestle. According to the folklore of China, this companion of the lunar goddess Chang'e is eternally pounding the elixir of life for his mistress or grinding the ingredients to make mortal medicines, while in Japanese and Korean versions of the tale, the rabbit pounds the ingredients of rice cake. Appropriately, all of China's lunar space probes from 2007 onwards have been named after Chang'e, and the unexpectedly long-lived lunar rover on the *Chang'e III* mission was named *Yutu*.

The very notion that there is a man in – or of – the Moon continues to play upon the imagination from a young age as human features appear to be writ large across its surface. However, there is a rather unlovely tale explaining who the man might be. It seems to have its roots in the Old Testament (Numbers 15: 32–36), when a man found gathering sticks on the Sabbath is put to death by stoning. A later German version relocates the events and has the wood-gatherer banished to the Moon:

> *Ages ago there went one Sunday an old man into the woods to hew sticks. He cut a faggot and slung it on a stout staff, cast it over his shoulder, and began to trudge home with his burthen. On his way he met a handsome man in Sunday suit, walking towards the church. The man stopped, and asked the faggot-bearer, 'Do you know that this is Sunday on earth, when all must rest from their labours?' 'Sunday on earth, or Monday in heaven, it's all one to me!' laughed the woodcutter. 'Then bear your bundle for ever!' answered the stranger. 'And as you value not Sunday on earth, yours shall be a perpetual moon-day in heaven; you shall stand for eternity in the moon, a warning to all Sabbath-breakers'. Thereupon the stranger vanished, and the man was caught up with his staff and faggot into the moon, where he stands yet.*[4]

Variations on the general theme appear to have promulgated across Europe, with the miscreant sometimes joined by a female companion – a Sabbath butter-churner – or offered a choice of punishment: immolation in the Sun or freezing on the Moon. An elegant painted scene can still be found in the chancel roof of St Benedict's Church, Gyffin, North Wales, which shows the man in the Moon, complete with his bundle of kindling.

The dramatic Moon

Interestingly, the concept of the Moon's inhabitant bearing a bundle of sticks appears in Shakespeare's depiction of pre-Christian, Classical Athens in *A Midsummer Night's Dream* (1595/6):

> *STARVELING [as MOONSHINE]: All that I have to say, is, to tell you that the lanthorn is the moon, I the man i'th'moon; this thorn-bush, my thorn-bush; and this dog, my dog.*[5]

Here the Sabbath-breaker carries not just any sticks, but thorns, which gave Christ his crucifixion crown, and so Shakespeare is either alluding to a double blasphemy or a further variant of the tale in which the miscreant strews the road to church with thorns to prevent his neighbours from attending mass. Irrespective of his intention, the playwright achieves a fascinating blend of ancient and contemporary that is as seamlessly complex as the chronologically tortuous mythological representations of Ancient Egypt.

Although we often associate the Full Moon with witchcraft and magical rituals on blasted heaths, there appears little real connection, save for the practicalities involved in enacting rites under a moonlit sky prior to artificial lighting. Nevertheless, the belief that witches had power over the Moon has long held sway, and the witches of Thessaly appear to have had particular potency in this regard. They make a comic appearance in Aristophanes's satire, *The Clouds* (423 BCE):

> *STREPSIADES: Suppose I bought a Thessalian slave, a witch, and got her to draw down the moon one night, and then put it in a box like they do with mirrors and kept a close watch on it.*
> *SOCRATES: What good would that do you?*
> *STREPSIADES: Well, if the moon never rises, I never pay any interest.*
> *SOCRATES: Why not?*
> *STREPSIADES: Why not? Because its reckoned by the month of course.*[6]

They also appear, more knowingly, in Plutarch's *Moralia* (100 CE), when he comments that the leader of the Thessalian witches, 'Aglaonice, the daughter of Hegetor, being skilful in astrology, made the vulgar believe, whenever the moon was eclipsed, that by means of some charms and enchantments she brought

Ivory netsuke figurine of Yutu, c.1701–1900

it down from heaven'.[7] Although these lunar powers appear determinedly ancient, in *The Tempest* (1610/11) Shakespeare, a writer well acquainted with Plutarch, would endow Caliban's off-stage mother, the Algerian witch Sycorax, with similar abilities. Prospero describes her as 'one so strong that [she] could control the moon, make flows and ebbs, and deal in her command without her power'.[8] Interestingly, Sycorax's lunar relationship is frequently reiterated, as her son is repeatedly described as 'mooncalf': formerly a fleshy mass in the womb, believed to have been created by a baleful lunar influence. In his usual fashion, Shakespeare developed the word, expanding its meaning to further encompass deformed or 'monstrous' creatures such as Caliban.

The lure of the Moon

For millennia, there has been a widespread belief in the association between the Full Moon and extremes of behaviour linked with mental illness. The very word 'lunatic', now thankfully outmoded, is an indicator of the former pervasiveness of this conviction among the general public and medical professionals alike. Both Pliny the Elder and Aristotle (384–322 BCE) believed that the brain's high water content made it susceptible to the tidal influences of the Moon. Research of 37 individual cases, published in the *Psychological Bulletin* of 1985, suggested that this was mere fallacy and the result of illusory correlation, whereby any odd events taking place during a noteworthy event or period, such as the rising of the Full Moon, might acquire greater significance.[9] However, a recent article by psychiatrist Charles Raison conjectured that the bright light of the Full Moon might produce sleeping disorders in those with mental illness, living on the streets or in generally anarchic personal circumstances, which could 'trigger erratic behaviour' in those affected by bi-polar disorders. Raison theorises that the ubiquity of electric lighting in the modern world has substantially lessened the effect on those who may once have been affected by the Moon's light, describing the phenomenon as a 'cultural fossil'.[10]

Many and varied are the tales of faerie-folk dancing by the light of the Full Moon. Folklorist Lady Jane 'Speranza' Wilde (mother of Oscar Wilde) collected many in her 1887 publication *Ancient Legends, Mystic Charms, and Superstitions of Ireland*, which has a special emphasis on the Moon and moonlight. Her faeries are 'the fallen angels who were cast down by the Lord God out of heaven for their sinful pride'. But she indicates that the 'fairies of the earth and the sea are mostly gentle and beautiful creatures, who will do no harm if they are let alone, and allowed to dance on the fairy raths in the moonlight'.[11] Their supernatural powers seem to be at their strongest with the rising of the Moon. The tale of Shaun-Mor of Innis-Sark, for instance, tells of an old man engaged as a porter for the faerie-folk whose increasing arrogance is mischievously repaid when they have an eagle deposit him on the Moon, the realm of the dead. He is greeted there by a pallid, scythe-wielding guardian. Intending no long-term harm, however, the faeries return Shaun to his home, where his wife takes his stories for drunken hallucinations.

Tales of faerie-folk and the draw of the Full Moon are not, of course, peculiar to Ireland, and there is a substantial tradition of similar creatures throughout the British Isles. It is worth mentioning the Selkie, a creature known in the highlands and islands of Scotland, particularly in the areas of the Orkneys and the west coast. Looking like seals bobbing and playing around the Caledonian coastline by day, Selkies can shed their sealskins and take the form of beautiful humans to visit land and dance on the beaches during the nights of the Full Moon. There is some debate about when Selkies have this ability, although most accounts confirm that the light of the Full Moon is a powerful draw. Essentially gentle creatures, lacking the dangerous traits of sirens or mermaids, Selkies feature in invariably tragic folktales and ballads – if they lose their skins, the beautiful creatures are compelled to remain on land as humans.

When I was first asked to write this piece, many friends nodded and intoned sagely, 'Ah, yes, werewolves'. I'm unsure what this says about me or, indeed, my choice of friends, but the inclusion of lycanthropy as a consequence of the Full Moon is more complex than one might expect. In Petronius's comic novel, *The Satyricon* (first century CE) a number of tales are related at the pretentiously vulgar dinner party of the freedman Trimalchio, one of which is the earliest literary example of a man transforming into a wolf. He strips naked before the narrator, Niceros, in a cemetery late at night and, although the Moon is said to shine like high noon, the moonlight appears to have little effect upon the miraculous transformation, save for allowing the narrator a clear view of the phenomenon. The Moon's phase remains unspecified. Similarly, Gervaise of Tilbury's *Otia Imperialia* (*c.*1211) makes no mention of the Full Moon when he writes, 'in England we have often seen men change into wolves, according to the phases of the moon. The Gauls call men of this kind gerulfi, while the English name for them is werewolves'.[12]

Historical tales of werewolves and their subsequent trials make no explicit mention of the Moon and many of the creatures are either said to have been cursed from birth or their transformations are simply beyond explanation. It is not until the release of Universal Studio's *The Wolf Man* (George Waggner, USA, 1941) that direct correlations are drawn between lycanthropy and the Full Moon. Curt Siodmak's script effectively constructs and presents much of the lore now commonly associated with the werewolf, including the idea of it being a contagious malady akin to vampirism. However, his rhyme, 'Even a man who is pure of heart and says his prayers by night, may become a wolf when the wolfsbane blooms and the autumn moon is bright', which is recited several times, has a lyrically authentic ring that appears to have created an aura of folkloric verisimilitude. With the film's sequel, *Frankenstein Meets the Wolf Man* (Roy William Neill, USA, 1943), Siodmak tweaked his verse to 'the moon is *full* and bright', increasing the association with the Full Moon. Seasonality is replaced by lunar phases and history, or rather folklore, is made. The epitome of the werewolf film is, to my mind, John Landis's witty homage to the sub-genre, *An American Werewolf in London* (USA/UK, 1981), which reiterates and refines Siodmak's ideas, placing the Full Moon at the centre of the legend, through script, *mise-en-scène* and, crucially, soundtrack. The film features no fewer than three versions of the popular song, 'Blue Moon', in addition to both 'Moondance' and 'Bad Moon Rising'.

Since 1941, numerous lycanthropic creatures have loped and scuttled across cinema screens at the command of the Full Moon, not all of them lupine. One of the most elegant and interesting appears in Hammer's dark fairy tale, *The Gorgon* (Terence Fisher, UK, 1964). Replacing the werewolf with the spirit of an ancient Gorgon, which inhabits the body of a young amnesiac woman in early twentieth-century Mittel-Europe, the film conflates the Gorgon myth with the Moon in a manner that turns out to be serendipitous in its relation to classical imagery. While faces normally appear in profile in the two-dimensional art of ancient Greece, the Gorgon is invariably depicted front-on with eyes staring widely, tongue lolling and tusks champing. One suggestion for this distinctive depiction is that the Gorgon is the Moon's shadow, devouring the Sun during a solar eclipse. In this interpretation, her painted face represents the 'other' overturning the natural order, creating terror and dismay.

While relations between vampires and the Moon may

○ *Belgian poster for* The Gorgon *(1964)*

not seem obvious, Bram Stoker's magisterial Gothic novel *Dracula* (1897) proposes that the undead can change their forms and 'come on moonlight rays as elemental dust',[13] and earlier in the novel, Stoker's young hero, Jonathan Harker, witnesses the Count's three vampire 'brides' materialising in just such a manner. Although undoubtedly utilised for atmospheric effect, the Moon is referenced no less than 79 times in *Dracula*; Stoker, amateur folklorist and friend to Lady Wilde, was evidently anxious to incorporate a degree of lunar lore within his *magnum opus*. We've already seen that there are Romanian traditions of vampires nibbling the Moon, and even the Sun, during eclipses, but there is also a tradition that the power of the human vampire – rather different from Stoker's vigorous and self-possessed vampire Count – is cyclical, being strongest at the time of the New Moon and waning with the Moon.

I would be remiss were I not to mention – or, rather, dispel – the myth of the Moon being made of cheese. It is an oft-recounted expression in many cultures, which has no basis in any genuinely held belief, mythological or otherwise, but refers instead to the naivety of an individual who might believe such a thing to be true. On occasions, the cheese in question is 'green', referring not to its colour but to its immaturity, perhaps like the individual under discussion. One might, therefore, suggest 'cream cheese' as a more apposite rendering.

In order to explain the title of this chapter, I would like to mention a superstition about pointing directly at the Moon. It was once believed in Western Europe that if one pointed at the Moon nine times – it is unclear whether this was at one time or during one's life – one would never succeed in entering heaven. Interestingly, a similar tradition has developed in modern Chinese culture, wherein pointing at the Moon offends the goddess, Chang'e, who will attempt to slice off your ear as you sleep, leaving behind a slight nick or papercut. This fanciful idea would appear to be of relatively recent origin and is probably best described as an urban legend. As such, it is evident that, even in the technologically advanced modern world, the Moon is still a potent and continuing source of fascination for the human imagination. As we learn more about the scientific realities of our closest celestial neighbour, one can only guess at the examples of lunar lore that may proliferate in the future throughout the cultures of our increasingly globalised world. After all, if there's one thing that humanity has always appreciated it is a good tale, well told.

○ *Black-figure terracotta stand with Gorgon by Kleitias (painter) and Ergotimos (potter), c.570* BCE

The Moon in China

Christopher Cullen

Gazes on the Moon

The Moon looks different from different times and places, but it also looks the same. In many parts of the ancient world it was a manifestation of a divinity; in China[1] it was, on the whole, a place rather than a person, although divine beings might inhabit it. This essay will leave aside the stories of such beings, how they got to the Moon, and what they did there, nor will it say anything about the poetic or romantic value that Chinese poets (like poets everywhere) found in the Moon and the light it casts. Instead, it will concentrate on the more practical aspects of the Moon as seen by people on Earth – whether emperors, officials, merchants or farmers – who had to organise their lives and set a date for future activities. Doing that needs a calendar. But what kind of calendar was it, and how did it involve the Moon?

The calendar

The earliest records of Chinese dates available to us come from inscribed animal bones, used by royal diviners in the late second millennium BCE, under the Shang 商 dynasty. These suggest that, at that time, the calendar in use was already fully lunisolar. That means that it used three time units, of which the first two depended on the Sun and the third on the Moon.

The first was the cycle of light and dark, commonly called the day. Days could, in effect, be numbered in a cycle of 60 using a two-character name formed by permuting cycles of ten and twelve characters in regular order. This cycle ran continuously, without reference to months or years, and appears to have run in unbroken sequence from very early on. The second was the year, a long-term cycle consisting of a whole number of days that follows the cycle of the seasons. Years were normally designated by their place in the reign of a ruler.

The third time unit was the lunar month, a medium-term cycle, consisting of a whole number of days, that attempts to follow the cycle of the Moon's phases, with the first day of the month *shuo* 朔 ('beginning') ideally containing the moment of conjunction of Sun and Moon (New Moon), when the Moon is not visible. Because the Moon's phase cycle (synodic month) takes on average just over 29·5 days (and any given cycle can vary from that value by up to seven hours), lunar months need to alternate between 29 and 30 days in length, with occasional pairs of successive long months, each of 30 days, to ensure that the months keep step with the Moon's phases. Lunar months were typically numbered rather than named, although the first month of the year was referred to as the 'standard month' *zhengyue* 正月. For most of Chinese history the month count in common use (the so-called Xia 夏 dynasty count) placed the standard month in early spring. It is the beginning of that lunar month that is commonly referred to today as 'Chinese New Year'.

But there was a further constraint: the year used for civil dating consisted not only of a whole number of days but also of a whole number of lunar months. Now, twelve lunar months last close to 354 days, whereas the cycle of the seasons repeats with a period of about 365·25 days (the tropical year). Thus, a year of twelve lunar months will fall 11 days short of the seasonal cycle, and after three years the year will be 33 days in advance of the seasons. To get back in step, an extra thirteenth month will have to be added to the year. This month was inserted between two normally numbered months and did not count in the normal sequence. It was designated as a *runyue* 闰月 or 'intercalary month'.

We can trace the work of Chinese specialists in calendrical science by checking the dates found in historical records. One famous such work is the *Chunqiu* 春秋 ('spring and autumn [annals]'), a terse annual chronicle of Confucius's home state of Lu 鲁 in modern Shandong 山东 province, running from the eighth to the fifth centuries BCE. This work contains a number of records of solar eclipses, which can only occur at the time of conjunction of Sun and Moon. The fact that in most cases when full data are given the eclipse is in fact recorded as falling on the first day of the month, *shuo*, suggests that the calendar makers were succeeding in tracking the phases of the real Moon. What is more, comparison of the sequence of days of given months in the 60-day cycle from one year to the next can reveal the presence of intercalary months. It thus appears that by the sixth century BCE a regular pattern of intercalations, seven in 19 years, had become standard practice. Taking 365·25

In this imaginary European depiction, Jesuit missionaries display their astronomy skills to the Shunzhi emperor (r. 1644–61) (dressed in red) who needs their assistance in creating a lunisolar calendar. Beauvais Manufactory, woven under the direction of Philippe Béhagle, c. 1697–1705.

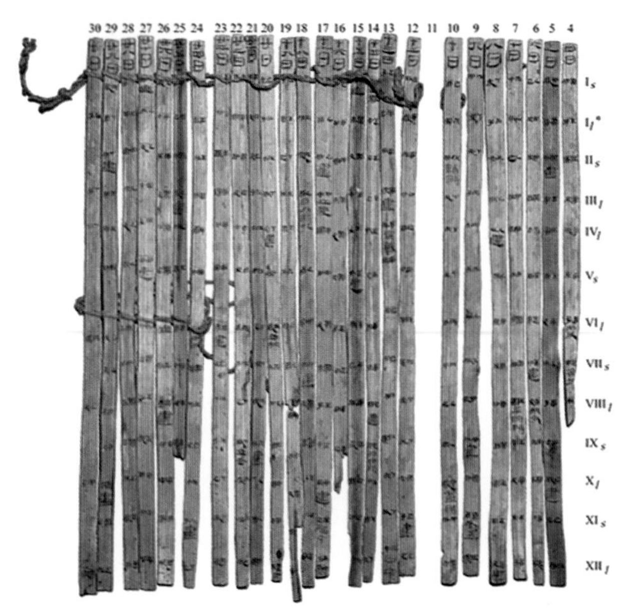

○ *Bamboo strip bundle type calendar, 69* BCE *(Han Dynasty)*

日轉度分	列衰	損益率	盈縮積	月行分
一日十四度十分	一退減	益二十二	盈初	三百七十六
二日十四度九分	二退減	益二十二	盈二十二	二百七十五
三日十四度七分	三退減	益十九	盈四十三	二百七十三

欽定四庫全書 晉書 十七

○ *Part of Liu Hong's lunar tables, showing variations in lunar speed in* Jinshu 晋书 *17, 17a,* Siku Quanshu 四库全书 *edition, c.1782*

days as the length of the Sun's annual cycle, the tropical year, we may calculate the implied length of the Moon's phase cycle, the synodic month, as follows:

> 19 tropical years contain $12 \times 19 + 7 = 235$ lunar months. Thus, 1 synodic month $= 19 \times 365{\cdot}25$ days $\div\ 235 = 29{\cdot}531$ days.

This is identical to the modern mean value to the same precision.

Eclipses and eclipse prediction

In the late third century BCE, the state of Qin 秦 succeeded in conquering all the other more-or-less independent states into which China had been divided for many centuries. This marked the beginning of what is commonly called the 'imperial age', during which the normal state of things was for most of the territory of the East Asian landmass to be ruled by a centralised government managed by an (in principle) non-hereditary civil service, ultimately responsible to an emperor who owed his throne to dynastic succession. All imperial governments maintained staffs of sky-watchers who observed and interpreted celestial omens, and calculators who enabled the emperor to confer on his subjects an accurate lunisolar calendar. From the first century CE onwards, we have detailed records of the successive astronomical systems, *li* 历, which these calculators developed to enable them to carry out the latter task.

Solar and lunar eclipses were matters of concern to omen interpreters and calculators alike. A solar eclipse was particularly worrying, because the darkening of the main light of the heavens and the symbol of the masculine cosmic principle, might, by analogy, suggest that the power of the emperor was under threat. Changes in the 'feminine' Moon were less alarming. By the first/second centuries CE, we begin to find references to explanations of eclipses in terms of the blocking of the Sun's light by the Moon in the case of a solar eclipse, or by the Earth in the case of a lunar eclipse, where the Earth was thought to cast a shadow on the Moon.

Independently of such explanation, it became known by the end of the first millennium BCE that the strong likelihood of a lunar eclipse could be predicted empirically by simple calculations based on a cycle of 23 lunar eclipses recurring every 135 months. Predicting a solar eclipse was a much more difficult task, because it depends on predicting the times when the Moon will line up directly between the observer and the Sun. Whereas the Sun simply appears to move relatively steadily around a fixed path in the sky, the ecliptic, the Moon moves at varying speeds around a path that crosses the ecliptic at an angle of about 6 degrees. To complicate matters further, the points at which it crosses the ecliptic (the nodes) themselves move round in a 19-year cycle.

It was not until the work of Jia Kui 贾逵 (30–101 CE) that the Moon's speed variation was understood. The full complexity of its path was first analysed by Liu Hong 刘洪 (129–210 CE), to whom the first successful prediction of a solar eclipse is attributed – probably the eclipse visible at Luoyang on 27 November 178 CE. From Liu Hong's time onwards, official astronomers began to make predictions of solar eclipses, although the reliability of such predictions was limited by the fact that the Chinese view of the cosmos saw the Earth as flat and extending across the celestial sphere, on the inside of which moved the celestial bodies, rather than as a spherical mass of diameter much smaller than the celestial sphere, as was the view in the ancient West. It was thus difficult to allow accurately for the changing points of view of observers at different positions on the Earth's surface. This remained a problem that had to be dealt with by a series of makeshift solutions until the Ptolemaic model of the universe was introduced to China in the seventeenth century.

The astronomical instruments of the imperial observatory on a section of the city wall of Beijing, as recorded in this print by Caspar Luyken, Amsterdam, 1698

○ Lunar Illumination in the Art of Africa

Christine Mullen Kreamer

> *Turn out the lights and watch the real ones in heaven –*
> *those our ancestors' imaginative minds used*
> *to mold a wonderful poetic imagery about themselves*
> *and their relation to the universe.*
> Anthony Aveni[1]

The words of anthropologist and cultural astronomer Anthony Aveni capture beautifully the wonder we all experience when looking up at the night sky. As he affirms, Africans have gazed upon the celestial firmament for millennia with equal intensity, making sense of the heavenly bodies and their movements as a navigational aid and to regulate agricultural and ritual calendars. African peoples and cultures have metaphorically brought the Sun, Moon and stars down to Earth as part of human culture, where they find expression through rich and complex social, political and religious systems and through the creation of works of art dating from ancient times to the present day.

The Moon – its presence in the night sky, its journey as it rises and sets, and its light that reveals or hides human activity – is a potent metaphor in the verbal and visual arts of Africa. A feminine symbol in many African societies, the Moon is often linked to life itself, through the regularity of lunar cycles that align with human and agricultural fertility and structure social and ritual activities. These are reckoned, in some regions, by the appearance of the Moon at certain phases against stars or star groups such as the Pleiades, Orion's belt and sword, and Sirius. The phases of the Moon are also central to the religious calendar of millions of African Muslims.

Africa's cultural and artistic connections with the Moon date back thousands of years. Ancient Egyptian and Nubian religious beliefs fostered a correlation between the major deities and the Sun, the Moon, the Earth, and with selected stars and planets known in Egypt during the time of the pharaohs. In the verbal and visual arts of Africa, including those of ancient Egypt, ideas about the Moon are often conceptually linked with the celestial body that dominates the sky during the day – the Sun. The Sun and Moon sometimes serve as metaphors for the complementary roles that men and women play

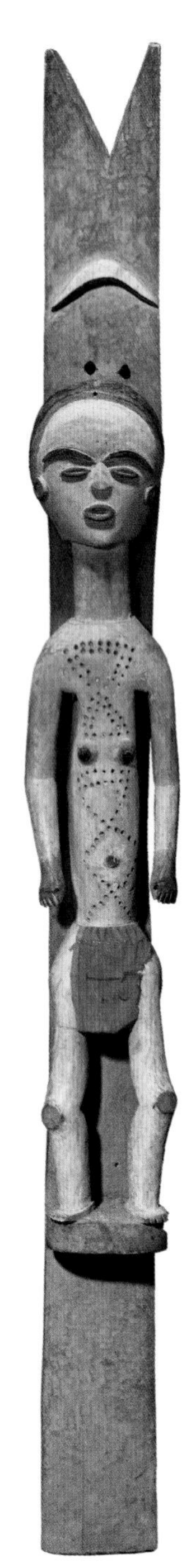

○ *Post figure (movenga), Tsogo or Sango artist, Gabon, early to mid-twentieth century*

Face mask by Nafana artist, Bondoukou region, Côte d'Ivoire, Ghana, mid-twentieth century

○ *Helmet mask (mwadi) by Sungu-Tetela or Tempa-Songye artist, Kasongo, Democratic Republic of the Congo, early twentieth century*

○ *Face mask (kifwebe) by Songye artistl, Democratic Republic of the Congo, early to mid-twentieth century*

in founding and sustaining community life. For example, Tsogo and Sango artists from Gabon carved paired male and female posts to ornament structures used by practitioners of the Bwiti spiritual belief system. A crescent moon typically ornamented the female post (see page 54), while a sun ornamented the male post. Together, these posts reinforced culturally-specific notions of community through initiation into Bwiti, through marriage and the creation of families, and through the ongoing presence of the ancestors in social life.

In parts of Africa, masquerade performances may be timed to lunar calendars or mark the phases of the Moon, with the Full Moon providing natural illumination for communities to gather outside for celebrations tied to planting and harvest seasons and for funeral ceremonies that mark the transition of the deceased into the ancestral realm. In northeastern Côte d'Ivoire and northwestern Ghana around the Bondoukou region, ceremonies involving the use of the Bedu masks of Nafana (see page 55) and neighbouring groups are performed according to a lunar cycle. The typically tall and brightly coloured plank masks appear in pairs, a smaller male mask often identified by crescent-shaped horns, and a larger female mask usually depicted with a disk superstructure. While the disk-shaped form of a Bedu mask might suggest its connections with the Full Moon, its primary lunar connection is in its performance cycle, with many performances occurring during a festival time referred to as the Bedu month, typically celebrated during late November when the moonlight is 'brilliant'.[2]

In central Africa, a visually dramatic mask collected by Methodist missionary John Noble White in 1924 was said to be used for the dance of the New Moon and for funerals and other occasions among the Sungu-Tetela or Tempa-Songye peoples of the Democratic Republic of the Congo. White's 1924 field photo documents the complete masquerade ensemble, which, in its original state, included the wooden mask and its fibre ruff, but also vulture and guinea fowl feathers that once projected from its crest and a fibre skirt and leopard skin costume that was worn by the masquerader. In performance, the mask, with its boldly painted striations and composite costume, would have towered impressively over the audience.

The mask's linear patterns link it visually to a wider group of striated masks called *kifwebe* (see opposite), some of which also have lunar connections. *Kifwebe* masking societies, found among the Songye and Luba peoples of the Democratic Republic of the Congo, served regulatory and judicial functions and were particularly associated with expressions of power and authority over potentially negative, anti-social behaviour. Dunja Hersak's research among the eastern Songye notes that *kifwebe* that were predominantly white were identified as female, with the colour white associated with moonlight and a host of positive attributes, such as 'goodness, purity, reproductive strength, peace and beauty'.[3] Male *kifwebe* masks, by contrast, are distinguished by boldly coloured bands of red, black and white.

Without specialised cultural knowledge, it might be difficult for many of us to see how the Moon is referenced in objects made by Luba and Tabwa artists from the Democratic Republic of the Congo, as there seem to be no obvious visual indicators. Yet scholarly field research conducted in the region in the 1970s and 80s by Mary (Polly) Nooter Roberts and Allen F. Roberts reveals, for example, that twins are called 'children of the Moon' and are linked with divination, the spirit world, and Luba kingship. This idea is referenced in a Luba headrest ornamented with a double figure motif suggesting twinned spirit mediums, linked arm-in-arm. 'Songs for twins' are sung by the attendants of diviners in order to 'call forth spirits' during consultations, but they are also sung on the night of the New Moon each month as well as for other important events such as funerals, initiations and the investiture of a new ruler. Beautifully carved single and double bowls (see page 58) are placed near the diviner during consultation as 'receptacles for moonlight' – the white chalk powder placed around the eyes of the diviner to provide intellectual enlightenment.[4]

This idea is also conveyed in the white-beaded strands and headdresses worn by Luba diviners and others charged with specialised knowledge. In the same region, quartz crystals, which were sometimes inserted into the top of the head of carved female figures, are said to capture and reflect moonlight, its radiance believed to inspire the dreams of diviners and rulers who owned such objects.

For the Tabwa, a Luba-related group residing in the southeastern Democratic Republic of the Congo along the shores and plateaux southwest of Lake Tanganyika, the Moon in its phases reflects the duality and complexities of the human condition, with the rising of the New Moon perceived as 'a sign of recognition, hope and rebirth'. By contrast, the disappearance of the Moon each month is inauspicious and a Tabwa metaphor for negative, anti-

Bowl by Luba artist, Democratic Republic of the Congo, late nineteenth to early twentieth century

Headrest attributed to the workshop of the Master of Mulongo, Luba artist, Democratic Republic of the Congo, mid to late nineteenth century

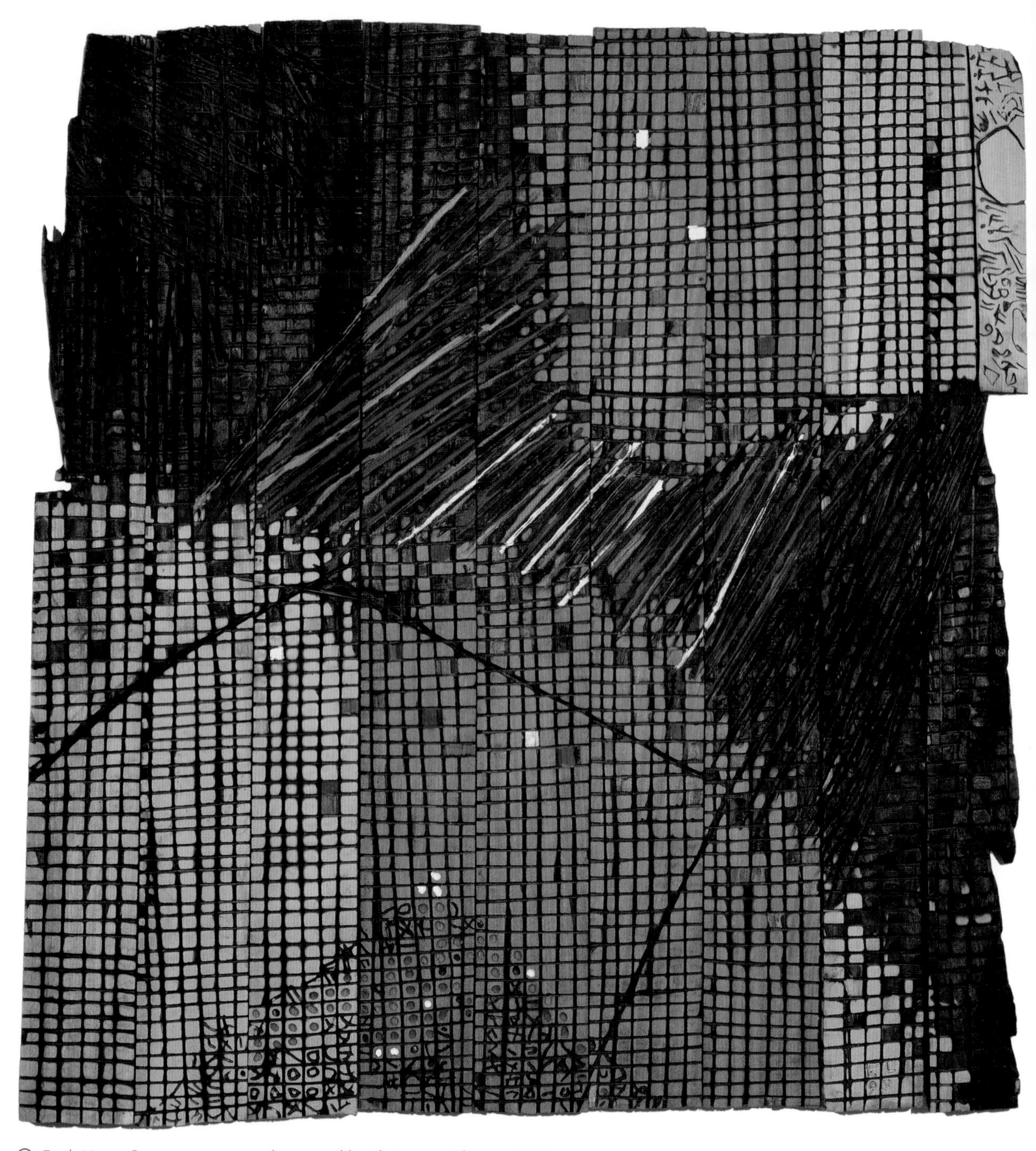

○ Earth-Moon Connexions, *painted on wood by El Anatsui, Ghana, 1993*

○ City in the Moon, *a woodcut print, by Adebisi Fabunmi, Ghana, 1960s*

social qualities. Tabwa figure carvings of ancestors bear scarification patterns that include parallel isosceles triangles called *balamwezi*, 'the rising of the new moon'.[5] According to Allen F. Roberts, the pattern 'unifies Tabwa artistic expression',[6] and it ornamented the human body (via painted or scarified patterns) as well as wooden neckrests and twin figures, (see page 59) plaited divination baskets, metal objects and beaded headgear and masks.

Contemporary artists from Africa also draw on the Moon as a source of inspiration. In *Earth-Moon Connexions*, (see page 60) artist El Anatsui (born 1944) captures something of the shimmering celestial bodies of our universe, while also suggesting astronomy's grid-like plotting that fixes stars, moons and planets within particular galaxies in deep space. Dots, lines and concentrated areas of colour suggest the movement and phases of the Moon over the landscape and its connection to seasonal calendars regulating agricultural work and the timing of rituals and other activities.

Adebisi Fabunmi's (born 1945) *City in the Moon* (see page 61) offers a more personal conception, presenting two linked zones defined by networks of tightly-spaced houses recalling life on Earth. While it may have been inspired by news of Western efforts to land a man on the Moon, the root-like structures that frame and contain this city, along with the curvilinear outlines that define its limits, have a subterranean feel. Might this be a thriving underground settlement that has escaped the notice of those on Earth – a hidden place that has thwarted the colonising tendencies of the West? Might the tightly spaced houses reference the urban sprawl of our own overcrowded cities, challenging Earth's leaders to seek alternative sites for human habitation? One is left to speculate on the artist's intentions, but such musings illustrate how works of art invite conjecture and resist singular interpretations.

The attenuated linear patterns of *Eclipse* (see opposite), an etching by Bruce Onobrakpeya (born 1932), communicate a sense of anxiety that has often been associated worldwide with unusual celestial phenomena, such as solar and lunar eclipses. The artist may have employed the metaphor of the eclipse, however, to convey societal unease in the face of actual or impending crises at the advent of the Nigerian Civil War (1967–70), also known as the Biafran Civil War, thus demonstrating, as have the other artworks discussed above, the relevance of celestial phenomena to social experience.

This was certainly the case in 1989 during my research among the Moba of northern Togo, when I attended the commemorative funeral rites of an important chief. The ceremonies coincided with the Full Moon, an auspicious time, full of promise and possibility, and a time of sufficient natural illumination for the all-night drumming and dancing that accompany a proper Moba farewell for a revered elder. My research assistants and I counted on the light of the Full Moon to guide us during our long walk back home to the village of Nano. Once the main funeral ceremonies concluded, we took our leave, and with the Moon rising behind us, we set off. Then something unusual happened: the light slowly diminished and, when we were too far along to turn back, we realised we were experiencing a full lunar eclipse. What was fascinating was the response to this celestial phenomenon. By the time we arrived in Nano, our neighbours were outside – young and old alike – drumming, banging on cooking pans, with the children singing over and over again, 'The Sun ate the Moon! The Sun ate the Moon!' The festive display had a serious purpose: it was designed to entice the Sun to give back the Moon it was devouring, and it went on until the Moon was seen again in its entirety. Only then was order achieved and the Moba universe restored to its proper balance.

○ Eclipse *by Bruce Onobrakpeya, Nigeria, 1967*

A Place that Exists Only in Moonlight: the Moon as Muse

Melanie Vandenbrouck

Katie Paterson, A place that exists only in moonlight, *2015*

In her conceptual series of Ideas (2015–), a series of haiku-like sentences, Scottish artist Katie Paterson (born 1981) invites us to imagine 'a place that exists only in moonlight'.

The sentence's poetry lies in its impossibility, an abstract idea that slips away as soon as you feel it within reach. Moonlight is transitory though regular in its patterns, ephemeral though cyclical. It has no shape, form or substance. It wanes and vanishes but we can rest safe in the knowledge that it will reappear. Elusive and fleeting, it never keeps its embrace for longer than it takes the Moon to cross our night-time sky. We share moonlight, yet none of us can grasp it. Or can we?

As sunlight (and to a lesser extent, earthlight) reflected on the Moon's surface and bounced back to Earth, moonlight's intensity depends on the lunar phases as much as on the whims of the weather. Just one millionth as bright as the Sun, the Full Moon's light reveals a world of softer edges, shadows and tonal range. As a phenomenon that affects our vision of the world and our emotions, the ethereal qualities of moonlight have long inspired artists. How moonlight wraps around a landscape to transform its colours and mood, the poetic interplay of darkness and light it creates, and how it evokes human passions, have all intrigued artists.

In the last years of his life, Peter Paul Rubens (1577–1640) was occupied with a nocturne painting, unusually executed without the help of his studio assistants. *Landscape by Moonlight* (1635–40) was inspired by Adam Elsheimer's night-time scenes and shares characteristics – the starry sky, defined constellations, the reflection of the Moon on

○ *Peter Paul Rubens,* Landscape by Moonlight, *1635–40*

water – with Elsheimer's *Flight into Egypt* (1609–10). Rubens had previously copied the painting, which features the first naturalistic night sky in Western art and was possibly influenced by the recent telescopic observations of Galileo. Rubens turned Elsheimer's religious subject and dramatic portrayal of the heavens into an introspective pastoral scene shrouded in a glow of such warm intensity that it looks like day rather than night. The overall effect supersedes naturalistic concerns: to unify a landscape under the radiance of the Moon would be the aspiration of many artists in Rubens' wake.

Rubens' nocturne was admired in Britain throughout the nineteenth century, notably by John Constable (1776–1837), and one senses its influence in his *Netley Abbey by Moonlight* (c.1833). Constable had visited the site, popular with poets and artists, with his beloved wife, Maria Bicknell, on honeymoon in 1816, and recorded his impressions in a series of pencil studies sketched from nature.[1] That he made the watercolour late in life when struck by inconsolable widowhood might account for the melancholy atmosphere, dominated by cool hues of blue. The scene is imbued with mystery and a sense of loss, emphasised by the lone figure contemplating a tombstone. The medieval monastery's ruins seem to be carved out of moonlight, while the trees' foliage picks up the Moon's glint. Like the Moon's silhouette these are the areas of the paper left untouched by watercolour wash, in a superb demonstration of Constable's power of suggestion. Like his contemporary J.M.W. Turner, Constable excelled in expressing the essence of a landscape through its sky and effects of light.

○ *J.M.W. Turner,* Moonlight on River, *c.1826*

In his Royal Academy debut in 1796, Turner (1775–1851) showed his debt to the eighteenth-century vogue for nocturnal marines in which the cool light of the Moon competes with warmer man-made light. His *Fishermen at Sea* sits within the romantic tradition of the 'sublime' in its depiction of humankind's subjugation to the forces of nature. Huddled together around a quivering lantern, the seamen are at the mercy of the elements. The theatrical moonlight emphasises the swell of the brooding sea, and though the Moon fleetingly safeguards the fishermen from treacherous rocks, as the clouds close in, so too does danger. Achieving naturalistic effects of moonlight on water was notoriously difficult, and with its sophisticated variations of light and virtuoso handling of paint, this oil painting heralds Turner as a master of atmospheric effects. It is in his sketches, however, that the qualities of moonlight assume a life of their own. The freer media of gouache and watercolour enable an immediacy and freshness of expression, and *Moonlight on River* (*c.*1826) (see opposite) is a lively impression of the transient brightness of the Moon. Turner conveys the physical sensation of light coruscating across the sky and glistening in the water below. Swiftly suggested in lead white gouache, the Moon and its reflection shine intensely on the blue wove paper Turner favoured in these years.

Moonlight not only guides the tone of a landscape, but may also convey the feeling of the subject represented and in turn affect the emotions of the beholder. Although Joseph Wright of Derby (1734–97) is better known as a painter of Enlightenment-era scientific demonstrations, *The Lady in Milton's 'Comus'* (1785) is distinguished by a romantic

○ *Joseph Wright of Derby,* The Lady in Milton's 'Comus', *1784/5*

○ *Tsukioka Yoshitoshi,* Genji yûgao no maki *('The Yûgao chapter from The Tale of Genji'), from* Tsuki hyakushi *('One Hundred Aspects of the Moon'), 1886*

sensibility. The scene depicts the moment when the Lady is separated from her brothers and lost in a wilderness in the thick of night. Hearing ominous sounds suggestive of her fate should the debauched Comus reach her, she finds brief respite in a moonlit clearing. Turning her eyes towards the source of light in hope, she invokes the Moon as a 'glistering Guardian, if need were /To keepe my life, and honour unassail'd'.[2] While Wright was interested in empirical observation and the depiction of natural phenomena, in this emotionally charged piece, the landscape is expressive of sentiment and pathos, the Moon's light creating a sense of heightened drama. Shrouding the figure in a soft halo, it lifts the darkness of both the Lady's surroundings and her predicament. Between sublime and sensibility, Wright's painting elicits the viewer's empathetic response to a virtuous woman in the throes of abject terror.

By contrast, Samuel Palmer (1805–81) imbued the Moon with a spiritual aura as it presided over his scenes of harvest, return from church and shepherds tending flocks. Moonstruck from a tender age, Palmer came to regard the Moon as a divine presence, in part inspired by John Milton, to whom he had been introduced by the equally visionary William Blake (1757–1827). Blake's illustrations of Dr Thornton's *Pastorals of Virgil* (1821) had also made an impression on Palmer's mind, and were perhaps still vivid when he illustrated his own translation from Virgil late in life. One of ten etchings for *An English Translation of Virgil's Eclogues*, completed by his son and published posthumously in 1883, shows a wizard and his assistants performing a ritual by an open sepulchre. Lit by flickering flames and the partially covered Full Moon, the mystery of the scene is emphasised by the chiaroscuro effects of a dense network of cross-hatchings. Palmer's fellow 'Ancient' and friend, Edward Calvert, was versed in paganism, and Palmer's image and translation – to which he added his own verses – bear pagan undertones suited to his unique blend of Christian mysticism.[3]

Contemplation of the Moon itself is a subject in Western and non-Western art alike. Caspar David Friedrich (1774–1840) conveyed the profound pursuit of gazing at the Moon, in several paintings of 'moonwatchers', in which his figures, seen from the back, invite us to identify with them and share their meditation. *Two Men Contemplating the Moon* (1819/20), of which he painted several versions, is considered one of the defining pictures of German romanticism, exemplifying human communion with nature. It shows two friends paused in their stroll as they behold a three-day-old Waxing Moon, its delicate crescent and faint earthshine gently illuminating the landscape in an amber glow that suits the figures' reverie.

In Japan, observing the Full Moon was and remains embedded in daily life and culture. Moon viewing platforms offer privileged viewpoints to enjoy its sight, and Zen gardens with raked and sculpted gravel capture its light. An important element of Japanese aesthetics, contemplating the Moon and moonlight falling on one's surroundings is considered a refined pleasure and source of inspiration for Japanese writers and artists. One example that combines word and image is the *Ehon Kyogetsubo* 絵本狂月坊 ('Moon-mad Monk', 1789), a privately published anthology of 72 *Kyōka* or humorous poems written by amateur poets by moonlight. Kitagawa Utamaro (1753–1806) provided five *Ukiyo-e* ('picture of the floating world') illustrations that admirably conjure people's infatuation with the silvery light of the Moon. The variety of viewpoints – coastal, urban, mountainous, rural and celestial – is unified by the recurring motif of the Full Moon.

Few artists better express Japan's intimate relationship with the Moon than Tsukioka Yoshitoshi (1839–92), with *Tsuki hyaku sugata* 月百姿 ('One Hundred Aspects of the Moon'), a series of *Ukiyo-e* woodblock prints, published between 1885 and 1892. It takes its inspiration from Japanese and Chinese myths, folklore, religion, literature and history, with the Moon as the common thread. The series unfolds as the portrayal of 100 characters and emotions harmonised with the Moon's phases and the distinctive qualities of its light, suggesting that the human psyche and condition are in tune with Earth's cosmic companion. One of the prints depicts the ghost of a young lady, killed by the jealous spirit of her lover's mistress. She stands, absorbed, amid her garden of 'moonflowers', her ethereal figure delineated by moonlight. A diffuse shade of blueish grey envelops the scene in cool sorrow. Her arrestingly pale profile, flowing hair, the sophisticated drapery, all contribute to the quiet elegance and ineffable sadness of the scene.

For most of human history, the Moon may have been a source of light, determining night-time activities and guiding our steps in darkness, but with the Industrial Revolution that relationship was to change dramatically. In our increasingly floodlit world, we have grown divorced from the spectacle overhead, and moonlight at its fullest is now confined to dark sky areas, like an endangered species in a conservation park. We may have lost sight of

○ *Darren Almond,* Fullmoon@Yesnaby, *2007*

moonlight, but Katie Paterson invites us to think its influence and unique properties afresh. Collaborating with scientists to measure its spectral intensity, she reproduced its effects through a room installation in which a lightbulb recreates the glow of the Full Moon. *Light Bulb to Simulate Moonlight* (2008) consists of 289 such bulbs providing 2,000 hours of illumination each, a lifetime's supply of moonlight. While the transformative powers of moonlight have long been part of folklore, Paterson's work is an invitation to pause and partake in a mindful experience of its embrace, and perhaps let our souls be healed from the stress, rush and isolation of modern life.

Another consideration of time and light is the *Fullmoon* series (1998–) by Darren Almond (born 1971). Shot under moonlight with long exposures of 12 to 30 minutes, it reveals the nocturnal landscape in ways the eye cannot see. *Fullmoon@Yesnaby* (2007) presents the dramatic coastline of Orkney Mainland, shaped by tectonics as much as by the tides. Yet it is the impossible daytime colours, obtained through the extended opening of the camera's shutter, that make this landscape seem so otherworldly. What the human eye sees, in the moonlit dark, as a monochrome world of blues, greys and silvers is an effect of our retina's poor perception of faint and diffuse light. Saturated with moonlight, the series presents a muted chromatic range and haziness that recalls the work of Turner and Friedrich, a lineage Almond readily acknowledges. With the ochre accents of the sandstone cliffs and the evergreen plateau, Almond's picture exposes a space that is neither day nor night, 'the light of the moon being amplified to cause the landscape to glow from within'.[4] Perhaps not unlike basking in the glow of Paterson's *Light Bulb*, Almond's experience of sightlessly shooting through dimly lit shadows, awaiting cloudless conditions, is one of mindfulness as he bonds with the moonlit landscape around him.[5]

By contrast, lunar eclipses remind us to look at light on the Moon itself. El Anatsui's lunar evocations are better known for his tapestry-like hangings made of thousands of bottle tops, where the metal's shimmering effects suggest its enveloping influence on our world at night, as in *Sacred Moon* (2007). Anatsui (born 1944) takes a new departure with *Eclipse Suite* (2016), a series of prints that can be read as a universal and personal exploration of time through its suggestion of the balletic interactions between Sun, Moon and Earth. The red orb in *Cadmium-Vermillion Eclipse* (see page 72) conjures the vision of the Blood Moon during a full lunar eclipse, when the Earth blocks the sunlight and the Moon becomes red or ruddy-brown from light reflected from the Earth's atmosphere. The deep, rich blue, achieved by the effects of the Sun on light-sensitive paper, may bring to mind the so-called Blue Moon, the second Full Moon in a calendar month.

Moonlight has excited the human imagination for millennia, but it would not be until the end of the twentieth century that light observed on the Moon's surface would be seen through an artist's eyes. Apollo 12 Lunar Module Pilot Alan Bean (1932–2018) was the fourth of 12 men to step on its dusty grey plains. In retirement, he sought to express on canvas what his fellow 'moonwalker' Edwin 'Buzz' Aldrin had described as the 'magnificent desolation' of,[6] to use space historian Andrew Chaikin's evocative words, a 'world on which only shadows move'.[7] *Home Sweet Home* (1983), one of Bean's early recollections in paint, shows the lunar module as his and Mission Commander Pete Conrad's vulnerable home-base on the Moon's barren terrains. The astronauts' footprints capture the rays of sunlight that bounce off the lunar regolith, but this is a stark, harsh, blinding light far removed from the subtle accents favoured in Japanese Zen gardens. Bean conveyed the unmistakably alien feel of the impenetrable darkness of the sky, rendered starless by sunlight brightly reflected on the lunar surface and causing his iris to contract. By his own account, Bean's practice evolved from 'paint[ing the Moon] exactly as I remembered it as an astronaut, and much the way it looks in photographs', to an approach more impressionistic in nature, reflecting that 'in an artist's vision, the way he or she chooses to see things is not the same as what he or she actually does...'. Inspired by Claude Monet's painted series, Bean reflected on how his painting 'has helped me understand that an artist's value to the world is to help all of us experience more fully, feel more connected with beauty, and become more completely human than we could without art'.[8]

We share moonlight, an elusive, transient, fleeting phenomenon that each of us has observed, but that none of us can truly grasp. Yet countless artists have shown that we can hold its infinite manifestations within the boundaries of a canvas, page or sheet, or in verses, musical scores and celluloid. Captured in this way, moonlight's power to transform the worlds around us and within becomes palpable. As Katie Paterson suggests, there is *a place that exists only in moonlight*. Perhaps this place is as precious, vast and unique as our own imagination.

Cadmium-Vermillion Eclipse, *an intaglio print, by El Anatsui, 2016*

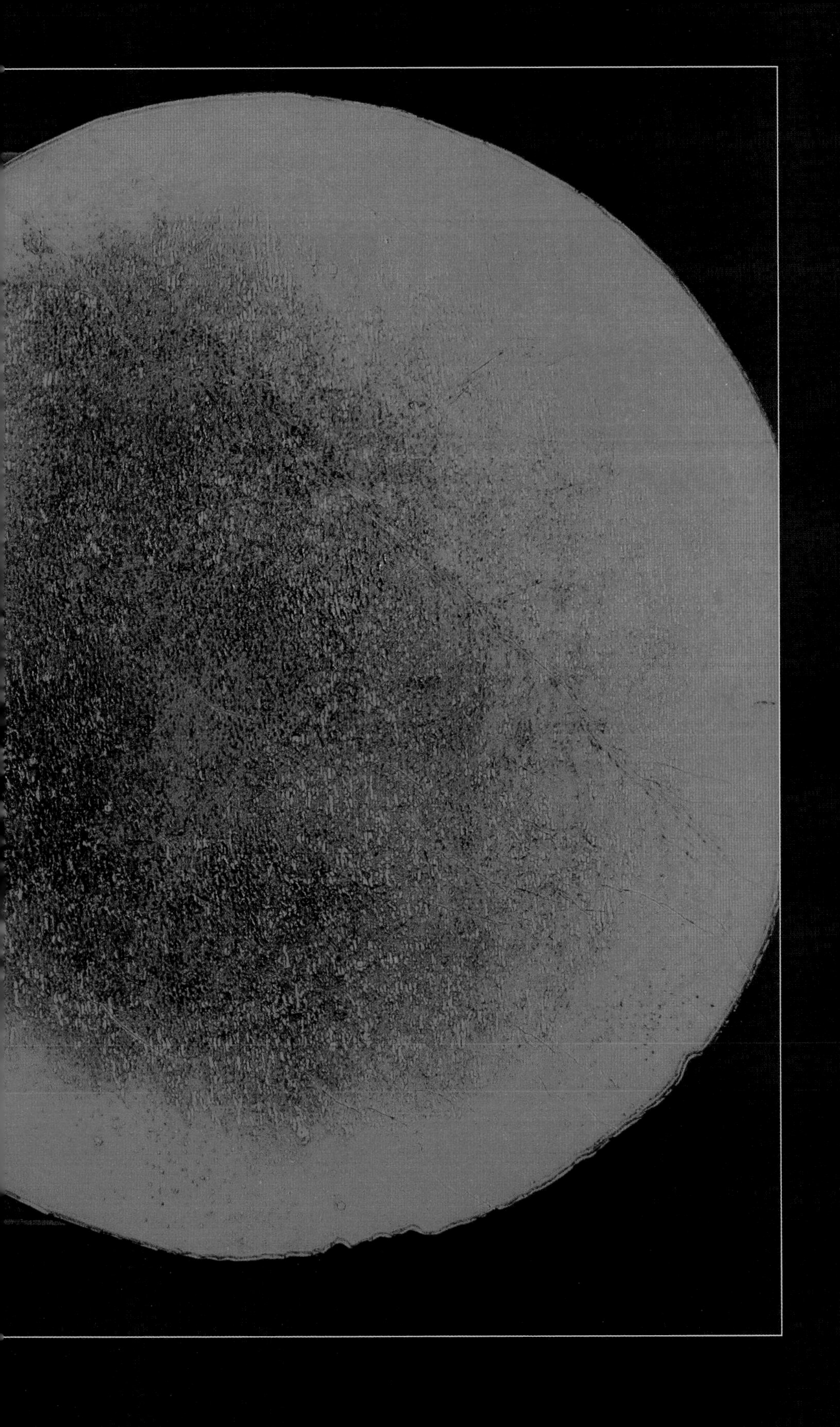

The Moon and Music

Scott Burnham

> *Et leur chanson se mêle au clair de lune,*
> *Au calme clair de lune triste et beau,*
> *Qui fait rêver les oiseaux dans les arbres*
> *Et sangloter d'extase les jets d'eau*
>
> *And their song mixes with the moonlight,*
> *With the sad and beautiful moonlight,*
> *Which makes the birds dream in the trees*
> *And streams of water sob with ecstasy*
>
> From Paul Verlaine, *Clair de lune* (1869)[1]

Just as music is a special kind of sound, moonlight is a special kind of light. Music and moonlight bathe the soul in mystery and enchantment, by suggesting emotional landscapes just beyond the realm of conscious thought.

Both Moon and music are famously mutable. The phases of the Moon are like the modes or keys of music, each producing different arrangements of light or sound. When the allegorical figure of Music steps onto the stage in the Prologue to Claudio Monteverdi's landmark opera *L'Orfeo* (1607), she speaks to the range of music's effects, claiming that she can calm agitated hearts but also inflame frozen hearts.[2] Just as the Moon pulls on the Earth's tides, music pulls on our emotional tides, from stirring anthems that can electrify entire nations, to intimate songs that seem to know what your own heart needs to hear.

Western musicians have long sought to invoke the power of the Moon, to claim kinship with its mysteries. The serenade by moonlight is a pervasive trope, and love songs of all manners have found their voice under the Moon. Such songs are powerfully concentrated in the art song tradition of the last two hundred years, and they are especially prevalent among nineteenth-century German *Lieder* set to poetry by that nation's finest poets.

Take Robert Schumann's 1840 setting of Heinrich Heine's *Lotosblume* ('Lotus Flower'). Struck shy by the Sun, this flower opens by the light of the Moon alone. Out of Heine's three-stanza poem, Schumann creates a small drama of love in the moonlight, using the simple expedient of repeated chords in the piano accompaniment, whose

L' ORFEO
FAVOLA IN MVSICA
DA CLAVDIO MONTEVERDI
RAPPRESENTATA IN MANTOVA
l'Anno 1607. & nouamente data in luce.
AL SERENISSIMO SIGNOR
D. FRANCESCO GONZAGA
Prencipe di Mantoua, & di Monferato, &c.

○ *Title page of Claudio Monteverdi's opera* L'Orfeo*, published in Venice in 1609 by Ricciardo Amadino*

harmonic changes register every emotional nuance. When the Moon comes out, Schumann changes key by means of a quietly alluring dissonance, as though to change the tonal lighting. The song then begins to speed up, the singer's voice to rise, as the flower opens, becomes fragrant, glows, weeps, and trembles with love – but also with the pain of love: Heine's last word is *Liebesweh* ('lovelorn').

Franz Schubert, the most celebrated *Lieder* composer of all, set Johann Wolfgang von Goethe's 1778 *An den Mond* ('To the Moon') twice, in 1815 and 1816. In Goethe's poem, the soul of the narrator is like Heine's lotus flower: it too opens up only in moonlight.

○ Spring Moon, *Alfred James Dewey, published by Reinthal & Newman, c.1914*

Again you fill bush and valley
With hushed and misty glow,
And also, once and for all,
You completely release my soul.[3]

Schubert's best-known setting of this poem (D. 296) matches the Moon's 'hushed and misty glow' by beginning not with a stable harmony but with a half-diminished seventh chord, one of tonal music's most bewitching sounds (its most famous instantiation is as the so-called 'Tristan' chord from Wagner's opera).

Perhaps the best-known Moon scene in the world of opera is Antonin Dvorak's stunningly beautiful 'Song to the Moon', from the first act of his 1901 opera *Rusalka*. The title character, a water nymph that has fallen in love with a human and wishes to become mortal, treats the Moon as her confidant, asking where her beloved is. Before Rusalka sings a note, the moonlight is rendered sonically by a distant horn sighing over muted strings, with oboes and clarinets each greeting Rusalka with breeze-like melodies that settle into a gentle oscillation in the strings as she begins her address to the Moon. At the almost unbelievably moving climax of her song, she implores the Moon to 'light up his faraway place, and tell him that I love him'.

A much cooler, more detached kind of moonlight is on display in Arnold Schoenberg's *Pierrot Lunaire* of 1912, a melodrama in the form of a song cycle for chamber ensemble and voice. Schoenberg set Albert Giraud's cycle of twenty-one poems merging the Moon with the characters of *commedia dell'arte*. This landmark work of twentieth-century art music brilliantly combines a cabaret spirit with many of Schoenberg's most progressive musical traits, such as the use of *Sprechstimme* ('speech song') and atonal harmony, and also with much older contrapuntal forms such as canon, passacaglia and fugue. Listen to any number of this cycle directly after Dvorak's love song and you will experience a disorienting plunge into the archly cynical, blackly humorous ethos of pre-war Vienna.

Moving out of the realm of art music and into that of popular American songs of the mid-twentieth century, we immediately come across Glenn Miller's 'Moonlight Serenade' of 1939, whose relaxed longing seems to sum up an entire age. Mitchell Parish added lyrics, taking rare advantage of Miller's famous melody: 'I stand at your gate and the song that I sing is of moonlight'. Each of the first four accented words (stand, gate, song, sing) in Parish's line matches with the repeated melodic figure in Miller's melody, which then climbs chromatically into the word 'moonlight'. The result is instantly recognisable as the enduring soundtrack for moonlit romance. A few later songs can also vie for this honour, such as Henry Mancini's 1961 'Moon River' and Bart Howard's 1954 'Fly Me to the Moon'. Neither song has much to do with the Moon per se, but neither suffers from the association.

Rock music, priding itself on its rebellion from the popular songs of earlier eras, is nonetheless heavily populated by songs invoking the Moon. One of the first and best of these is 'Bad Moon Rising', released in 1969 by Creedence Clearwater Revival. In this case, the Moon is a harbinger of disaster ('Don't go 'round tonight, it's bound to take your life, there's a bad moon on the rise') in a song that perhaps taps into voodoo traditions associated with the bayou country of Louisiana. Other songs invite us to dance in the light of the Moon, such as Van Morrison's 'Moondance' (1970), and King Harvest's 'Dancing in the Moonlight' (1972). In Cat Stevens's beguiling 'Moonshadow' (1970), seeing one's own shadow in the moonlight suggests something like an entire philosophy of life, with an Eastern tinge.

But the Moon doesn't need words to insinuate itself into music. Two of the most beloved pieces of Western piano music are forever linked by the invocation of moonlight: Ludwig van Beethoven's 'Moonlight' Sonata and Claude Debussy's *Clair de lune*. Debussy's composition was published in 1905 as the third movement of his *Suite Bergamasque*, and it was inspired by Paul Verlaine's poem 'Clair de lune', quoted above as an epigraph to this chapter. But Beethoven had no idea that his 1801 'sonata quasi una fantasia' (Op. 27, No. 2) would be known by posterity as the 'Moonlight'. The name arose shortly after his death, when publisher Ludwig Rellstab associated the first movement with 'a boat visiting the wild places on Lake Lucerne by moonlight'.[4] Although the name has been ridiculed by musical purists as a publisher's gimmick, it continues to resonate, and for good reason.

The openings alone of these two movements instantly convey a telling mood. Yet they could hardly be more different: Beethoven's *Adagio sostenuto* features brooding arpeggios, slowly morphing harmonies, and a plaintive melody, while Debussy's gently spangled thirds in a higher register seem themselves to shed a kind of moonlight onto the scene. Each movement soon intensifies, Beethoven's with heart-piercing dissonances, Debussy's with a flowing

Pierrot admiring the Moon and the stars, 1906

'Moonlight Serenade' by Glenn Miller, released in 1939

suffusion of sixteenth notes that gradually subside back into the glimmering light of the opening. Both movements are readily heard as nocturnal, Beethoven seeming to improvise in a state of shadowy melancholy and Debussy seeming to capture the quality of moonlight itself, as well as its effects on the beings awash in its wistful illumination.

Finally, the Moon doesn't even need audible sound to maintain its association with the spirit of music. In the ancient metaphysical concept of the 'Music of the Spheres', the Moon, along with the Sun and the planets, was thought to participate in a higher kind of universal music, a divine harmony not perceptible by the human ear. This is to place the Earth, Sun and Moon in an eternal dance that is both harmonic and rhythmic, comprising perhaps the most exalted music of all. But for so many of us here on Earth, the Moon's most meaningful place in the world of Western music will surely remain its role as the nocturnal sun, whose mysterious light opens hearts and souls to the music of love.

Ludwig van Beethoven's 'Moonlight' autograph score

THROUGH THE LENS

In the early seventeenth century, putting two lenses in a tube to make a telescope changed how people thought about the Moon. Two centuries later, the camera promised to do the same. Both technologies offered a way of understanding and imagining a world that remained frustratingly out of reach.

But images are made and not found. What an observer sees and records has always been a complex matter of interpretation, rooted in their own assumptions and preoccupations. Whether philosophers, astronomers, artists, photographers or astronauts, lunar observers have had to work hard to define and interpret what their equipment showed them, and the images they subsequently produced. At the same time, it was through these images that others came to see the Moon as a world of its own.

Megan Barford

The Telescopic Revolution

Richard Dunn

The Moon became a new world in the seventeenth century, mainly due to a single invention, the telescope. First announced in the Netherlands in the autumn of 1608, the new optical aid spread across Europe in a matter of months, changing the way people looked at the world around them. As the first instrument to extend one of the human senses, the telescope's revolutionary promise was clear at once, and although it was promoted for terrestrial viewing, the possibility of using it to look to the sky was recognised just as quickly.

The first telescopic observations of the Moon were made in 1609 and rapidly changed how people thought about the universe. But what one saw depended on one's assumptions. This is clear in the work of the first telescopic astronomers. It was an eminent English mathematician and astronomer, Thomas Harriot (1560–1621), who produced the first datable telescopic lunar drawing. On 26 July 1609, he sketched the Moon as seen from Syon House near London with a telescope that magnified six times.

Harriot made further lunar observations and drawings over the next few years, as did others in Britain. In February 1610, Sir William Lower described to Harriot the Moon he saw from Wales using a 'cylinder' his friend had sent:

> *In the full she appears like a tart that my cooke made me last weeke; here a vaine of bright stuffe, and there of darke, and so confusedlie all over. I must confess I can see none of this without my cylinder.*[1]

Harriot's observations and drawings reflected his mathematical and cartographic interests, with light and dark patches shown like land and oceans on a chart.

It has been suggested that Harriot was interested in mapping the Moon to answer questions about its motion, especially to determine whether it 'wobbled' in its monthly orbit around Earth, a phenomenon known as lunar libration. His observations did indeed demonstrate libration but were not published in his lifetime, as this was not Harriot's way. Long secure in a position with a wealthy patron, the Earl of Northumberland, Harriot chose to communicate with his scientific contacts through letters and manuscripts. This is largely why he failed to achieve the fame accorded to a very different lunar observer.

The second person known to have drawn the Moon seen through a telescope had much more public ambitions. By May 1609, Galileo Galilei (1564–1642), a mathematics professor at the University of Padua, knew of the new invention and was making telescopes of his own. In autumn 1609 he looked to the night sky with a 20-powered *perspicillum*, as he called it. What he saw was so extraordinary, he believed, that he quickly published his findings in the *Sidereus Nuncius* (Starry Messenger) in March 1610.

Galileo's book caused a sensation, as he hoped: he wanted to attract the patronage of the Grand Duke of Tuscany and gain employment at the Florentine court. Sir Henry Wotton, Ambassador to Venice, immediately wrote to the English Secretary of State describing, 'the strangest piece of news ... ever yet received from any part of the

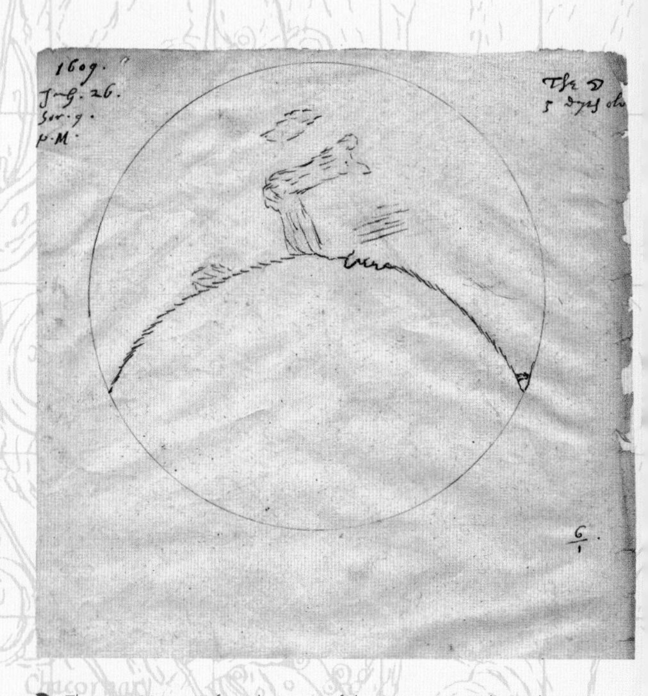

Thomas Harriot's first drawing of the Moon seen through a six-powered telescope, 26 July 1609

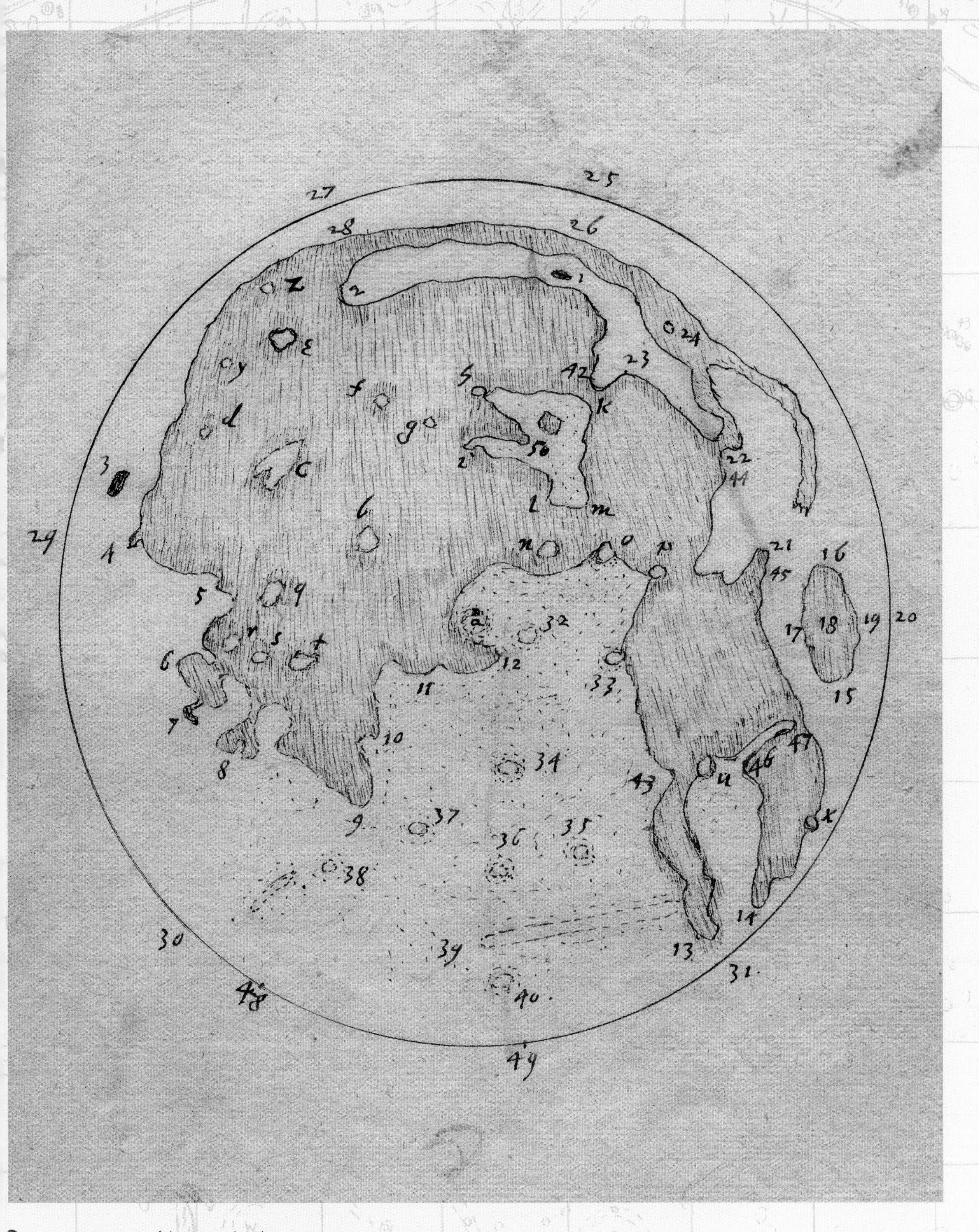

A composite map of the Moon by Thomas Harriot, probably drawn between 1610 and 1613

The iconic portrait of Galileo Galilei (1564–1642), painted by Justus Sustermans c.1639, shows the Italian polymath holding the instrument that made his name

OBSERVAT. SIDEREÆ

ſcatẽt:adeo vt ſi quis veterẽ Pythagoreorũ ſentẽtiam exſuſcitare velit, *Lunam ſcilicet eſſe quaſi Tellurem alterã, eius pars lucidior terrenã ſuperficiẽ, obſcurior vero aqueã magis congrue repræſentet:* mihi aũt dubiũ fuit nunquã *Terreſtris globi a lõge conſpecti, atq; a radiis Solarib. perfuſi, terreã ſuperficiẽ clariorẽ, obſcuriorẽ vero aqueam ſe ſe in conſpectũ daturã.* Depreſſiores inſuper in Luna cernuntur magnæ maculæ, quam clariores plagæ: in illa enim tam creſcente, quam decreſcente ſemper in lucis tenebrarumq; confinio, prominente hinc inde circa ipſas magnas maculas contermini lucidioris, veluti in deſcribendis figuris obſeruauimus, neq; depreſſiores tantũmodo ſunt dictarum macularũ termini, ſed æquabiliores, nec rugis, aut aſperitatib. interrupti. Lucidior vero pars maxime ꝓpe maculas eminet, adeo vt & ante quadraturã primã, & in ipſa ferme ſecunda circa maculã quãdã, ſuperiorẽ, borealẽ nẽpe Lunæ plagã, occupãtẽ valde attollantur tã ſupra illã, q̃ infra ingẽtes q̃dã eminẽtiæ, veluti appoſitæ ꝑſeferũt delineationes.

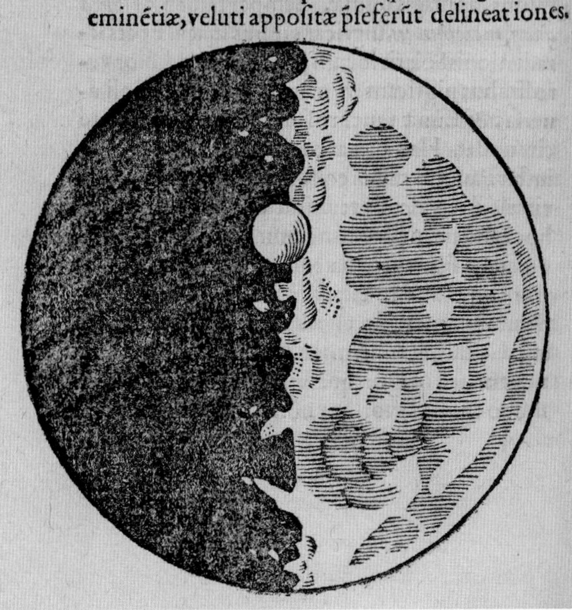

RECENS HABITÆ.

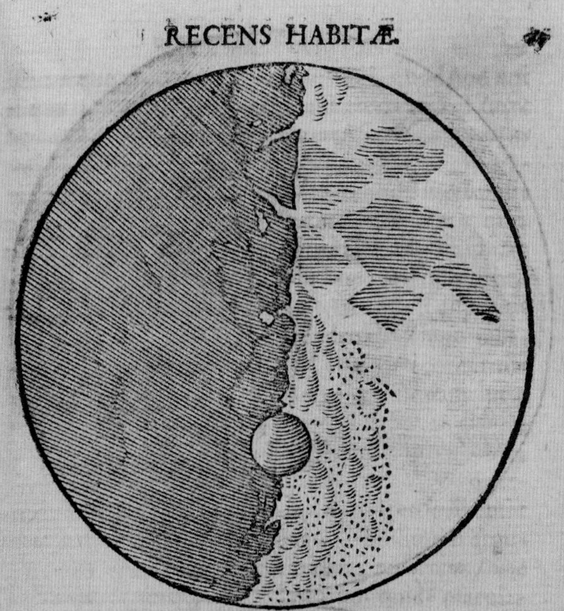

Hæc eadẽ macula ante ſecundam quadraturam nigrioribus quibuſdam terminis circumuallata conſpicitur, qui tanquam altiſſima montium iuga ex parte Soli auerſa obſcuriores apparent, qua vero Solem reſpiciunt, lucidiores exſtãt, cuius oppoſitum in cauitatibus accidit, quarum pars Soli auerſa ſplendens apparet, obſcura vero ac vmbroſa, quæ ex parte Solis ſita eſt. Imminuta deinde luminoſa ſuperficie, cum primũ tota ferme dicta macula tenebris eſt obducta, clariora montium dorſa eminenter tenebras ſcandunt. Hanc duplicem apparentiam ſequentes figuræ commonſtrant.

B

The Moon revealed by Galileo's telescope, from Sidereus Nuncius *(Venice, 1610)*

world'. Galileo was claiming that the Moon was 'not spherical, but endued with many prominences, and, which is of all the strangest, illuminated with the solar light by reflection from the body of the earth'.[2] This contradicted established doctrine that the planets, including the Moon, were perfect and unchanging, held in fixed spheres that rotated around the Earth. Galileo had discovered other moons too: Jupiter had four of its own, he announced. Earth's moon was no longer unique.

Galileo and Harriot saw and drew the Moon differently. Harriot's cartographic approach mapped the lunar features in two dimensions. Galileo emphasised the three dimensions of a lunar topography. By comparing a series of observations, Galileo noted how the Moon's appearance changed, which he interpreted as the movement of shadows across a rugged terrain. The Moon's surface was 'not unlike the face of the earth, relieved by chains of mountains and deep valleys', he wrote.[3] His interpretation drew on his artistic training in perspective and the rendering of light and shadow. William Lower described a similar change in perception. In the Moon, he wrote to Harriot, 'I had formerlie observed a strange spottednesse al over, but had no conceite that anie parte thereof might be shadowes'.[4] Reading the *Sidereus Nuncius* had transformed what his telescope revealed.

Galileo's extraordinary claims were not accepted by all. Some critics suggested that the Moon's appearance came from its non-uniform density and not its irregular surface. Others likened it to a crystal ball or a sunlit cloud, explaining the apparent imperfections as optical illusions. Opponents invoked the generally accepted view that the terrestrial and celestial regions were made of different matter. It did not follow that the telescope worked in the same way in both regions. 'On Earth it works miracles', wrote one, 'in the heavens it deceives'.[5] And even if one did accept that the telescope showed what was really out there, Galileo still had to explain why the Moon's edge seemed smooth, as it should look rugged if he were correct about its mountainous terrain.

Others embraced Galileo's new Moon. Italian artists began to use telescopes and to paint the 'telescopic Moon'. Within months of the publication of the *Sidereus Nuncius*, the artist Ludovico Cigoli included Galileo's discoveries in a painting for the Pauline (or Borghese) Chapel in the Basilica of Santa Maria Maggiore, Rome. He showed the Virgin Mary standing on a rugged, imperfect Moon.

The telescope's new discoveries opened up the possibility of life on the Moon as well, as Galileo had described an Earth-like environment. This set the stage for serious and satirical writings. In *Ignatius his Conclave* (1611), the metaphysical poet and preacher John Donne (1572–1631) imagined a telescopic device drawing the Moon earthwards as a vehicle for deporting the Jesuits he despised. Others speculated about life already there. Francis Godwin's *The Man in the Moone* (1638) was carried to an inhabited Moon by a flock of 'gansas' (wild swans or geese), and in the same year John Wilkins's *The Discovery of the World in the Moon* speculated seriously about life there. Both credited Galileo's telescopic discoveries. So too did the astronomer Johannes Kepler, who wrote that there were inhabitants not only on the Moon, but also on Jupiter, as the four moons Galileo had found could only have been created for Jovians, just as our Moon was for us.

While Galileo's interpretation of his lunar observations played up the topographical implications as a way of challenging physical conceptions of the universe, the cartographic emphasis of Harriot's work remained an important motive for other observers. Indeed, two of the most important, Giovanni Battista Riccioli (1598–1671) and Johannes Hevelius (1611–87), openly acknowledged the telescope as essential for the best Moon mapping. Working from his private, rooftop observatory in Danzig (Gdańsk), Hevelius produced some of the finest lunar maps of the age. The frontispiece to *Selenographia* (1647), his book of Moon observations, honoured Galileo, shown on the right with his telescope, and foregrounded the telescope as the main tool of lunar mapping. At the top of the frontispiece, the personification of Contemplatio (contemplation or survey) holds a telescope, with the Moon to the left and a quote below from Isaiah 40:26, 'Lift up your eyes on high, and behold who hath created these things'. The message was clear: the telescope had revolutionised how we saw the Moon.

The frontispiece to Johannes Heveluis's Selenographia (Gdańsk, 1647)

Mapping the Moon

Megan Barford

In 1955, Hugh Percy Wilkins's *Mysteries of Space and Time* described to readers the shock of looking at the Moon through a telescope. The amorphous patches of light and shade familiar to all who had gazed on Earth's satellite resolved themselves into awesome detail: towering peaks, gaping cracks, huge craters, deep pits, shattered rocks, all thrown into relief by the darkest of shadows. Wilkins (1896–1960), a Supply Ministry civil servant who observed the Moon from his garden in Bexleyheath, southeast London, was at this point one of the world's foremost lunar observers. He tempered his description, however, by noting the 'training of mind' that accompanied looking through a telescope. What Wilkins saw was 'an orb of mystery where strange things have taken place in the past and still seem to be in progress', and he added that 'those who bestow an occasional glance on the moon see nothing of this. To them, the moon is a world without change, a dead world and a museum piece, but to those people who observe the moon carefully whenever conditions are favourable, the wonder and the mystery of her fantastic landscapes grows continually.'[1]

Words on the wonder of looking at the Moon through a lens, the process of learning to see in a shifting technological environment, can be found from the invention of the telescope in the early seventeenth century on. As Richard Dunn emphasises in his chapter on the telescopic revolution, what people saw, and drew, was a complex matter of interpretation, based on the assumptions brought to the telescope and the notebook. This didn't end in the seventeenth century. As we will see, it continued to be true of different ways of understanding what was seen through different arrangements of lenses pointed at the Moon, as well as thinking about why one might look, draw, collate and share in the first place. The images of the Moon made by Galileo Galilei, and published in his *Sidereus Nuncius* in 1610, for instance, were not maps, but images of the Moon with its features hugely exaggerated. This was artistry, rather than flaw, and helped to emphasise Galileo's point. The images, which accompanied a remarkably detailed verbal description, were intended to accentuate the shadows thrown on its surface at different phases. Galileo was presenting the case for an uneven surface – for the mountains and craters that would become key components of selenography (literally 'moon drawing').[2]

Images, then, were argument, and in the developing field of selenography, this could work in different ways. Making an observation through a telescope was necessarily a private affair, as one eyepiece meant only one observer. As seventeenth-century observers looked and interpreted, they developed ways of demonstrating their authority, showing readers that what they had seen was trustworthy and reliable. One example of this is Polish astronomer Johannes Hevelius's *Selenographia* (1647). For the first half of the volume, Hevelius described the nature of his observations, not just writing about the equipment used to grind his lenses, but including images of it too. Picturing himself at his telescope, hand on hip, for example, showed just how stable his instrument was on its stand. In common with most accounts of natural philosophy in the seventeenth century, Hevelius's intensive description served to bring readers as close as possible to the actual observations whose results – in the form of highly developed Moon maps, diagrams and descriptions – were shown later in the book.

At the same time as astronomers were giving detailed accounts of their telescopes in order to give authority to their maps, telescope makers began to use impressive visualisations to endorse their work. Telescopes became business, and producing images of what could be seen through a specific instrument made in a particular workshop became a way to show its superiority. Neapolitan telescope maker Francesco Fontana (1580–1656) was one of the earliest to construct a telescope with two convex lenses. Although this inverted the image, it allowed for a wider field of view, and eventually became standard in telescopes for making celestial observations. Fontana circulated prints of the Moon made with his telescope, to demonstrate the virtues of his design, and of his lens-grinding technique. Eventually he published a well-illustrated book, one of the functions of which was the same: published observations that served not just as an illustration of the solar system, but also as an advertisement for the newest instrument of investigation.

As more observers turned their telescopes to the Moon, selenography became an important part of seventeenth-

Hevelius observing, in Johannes Hevelius's Selenographia *(1647)*

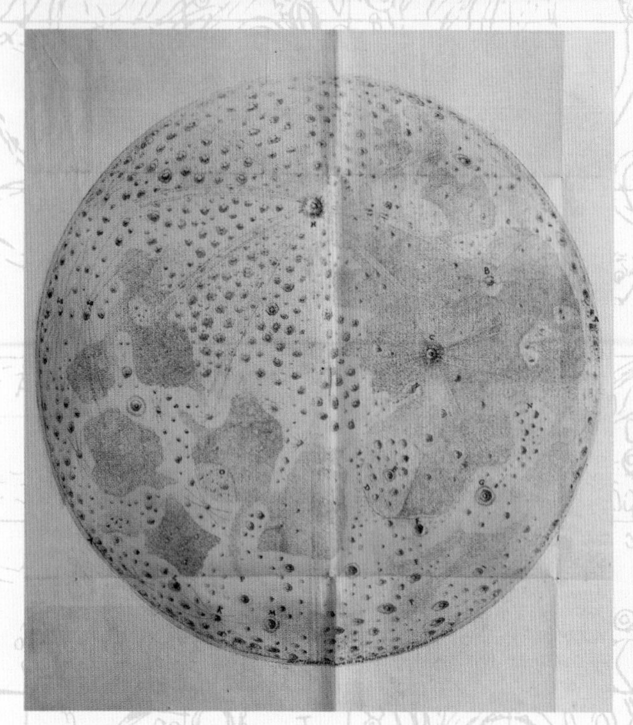

Moon map from Francesco Fontana,
Novae celestium terrestriumque rerum observationes *(1646)*

century astronomy. When the Jesuit scholar Giovanni Battista Riccioli produced his monumental 1,500-page work on the state of contemporary astronomy in 1651, it had to include selenography. Most significantly, this included a pair of Moon maps by Riccioli's assistant, Francesco Grimaldi, compiled from existing maps and new observations. Observing the Moon through the telescope was used to support new understandings of the place of the Earth in the universe. Astronomers, including Riccioli and Grimaldi, developed a further purpose: mapping the Moon to understand our place on Earth.

The idea that a map of the Moon could enable the calculation of longitude on Earth (how many degrees east or west one was from a given point) had a lot of currency in learned circles. The suggestion was that, during a lunar eclipse, observing the exact time that the edge of the shadow of the Earth passed over a specific spot on the Moon, compared with the exact time the same thing happened in a different location, would allow the calculation of longitude. Such an endeavour required detailed maps and a shared nomenclature, and it was for this reason that in 1645 Michael van Langren (1598–1675), part of a prominent family of Flemish cartographers in service to the Habsburg monarchy, published what is sometimes described as the first true Moon map: a representation that emphasised a schema of features over naturalistic representation, and emptied the lunar surface of the shadows given such prominence in the drawings of earlier observers such as Galileo.

The function of lunar maps, as instruments for geographical and astronomical investigation, was clear to those involved. Some even developed Moon maps specifically to be copied and drawn upon to enable them to function as instruments. One of the most surprising Moon maps of the seventeenth century is found towards the end of Hevelius's *Selenographia*. It is a template designed to help later astronomers record their own observations. Drawn without a border for the lunar disc, so that observations at any point in the Moon's libration could be recorded with ease, the template gives only the outlines of the main lunar features. The map was designed for plotting the passing of the shadow of the Earth over the Moon in a lunar eclipse, again to facilitate the calculation of terrestrial longitude.[3]

So evident was the link between the geography of Earth and that of the Moon to those involved in selenography, that towards the end of the seventeenth century, Jean-Dominique Cassini (1625–1712), Director of the Observatoire de Paris and Astronomer to Louis XIV, frankly rebuffed any criticism that selenography might, ultimately, be wasted labour: 'Those who do not look at things in depth, think they [maps of the Moon] are useless descriptions of an imaginary country. They are surprised that people of common sense enjoy themselves making such exact maps of a lunar world where certainly no-one will ever go, whether to make conquests or found colonies.' He went on to emphasise that Moon maps turned into instruments for establishing terrestrial longitude could further enable the long-distance imperial and trade networks in which European states were so heavily involved at the time. Indeed, he pointed out how detailed representations of the Moon 'serve to mark very precisely the places of the Earth, and to improve geographical and hydrographic maps, without which it is impossible to make long journeys and to trade with far off people'.[4] Where contemporaries used the motif of colonising the Moon as a rich resource for satire, astronomers used colonies on Earth as a reason for further selenographical work.

Those Moon maps of Grimaldi and Riccioli, and of Hevelius, remained the most used for most of the eighteenth century. Astronomers made Moon maps during the 1700s, but they largely remained unpublished, and therefore little

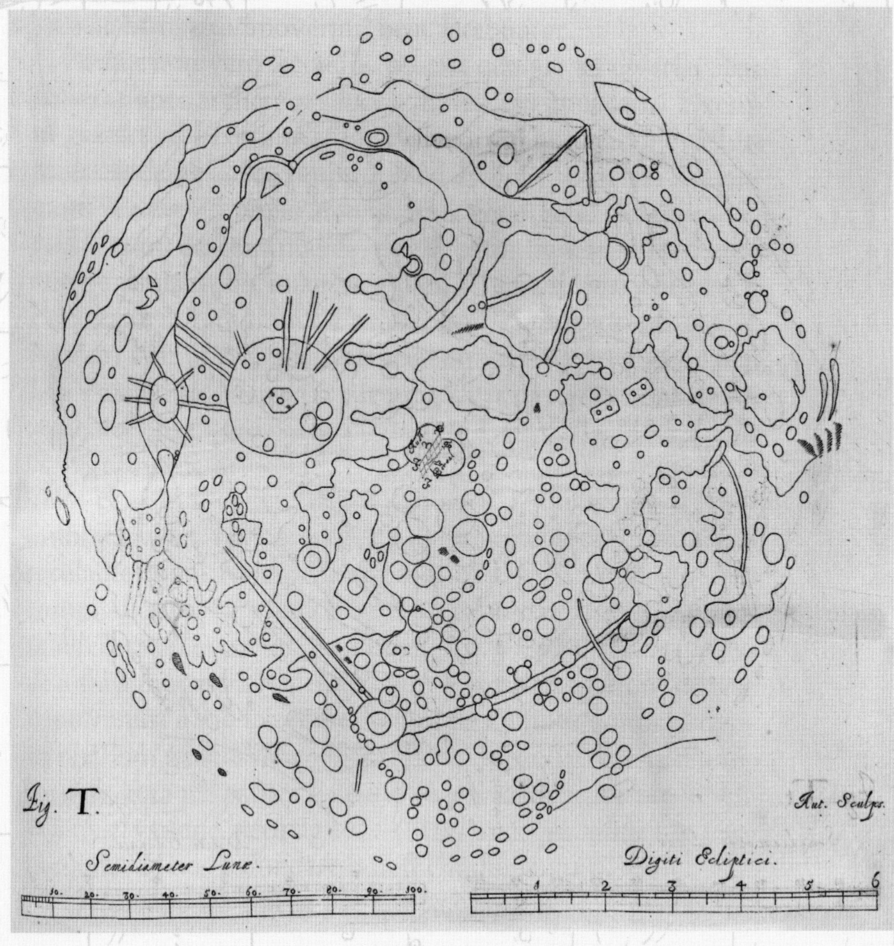

'Template' Moon in Johannes Hevelius's *Selenographia* (1647)

known. This collection of work included that of German cartographer and astronomer Tobias Mayer (1723–62), but his map was not published until after his death, and then at the back of a volume of various previously unpublished papers. Mayer had also begun to make a lunar globe, reasoning that calculations would be greatly facilitated by a three-dimensional model, something which earlier observers, including Hevelius, had also suggested.

The potential utility of a lunar globe, and the demonstration of astronomical, artisanal and artistic prowess that creating one would entail, meant that others explored the possibilities of such an instrument. Artist John Russell (1745–1806), whose evocative Moon pastels are discussed in the following chapter by Melanie Vandenbrouck, also turned the results of his intensive observations to this purpose. In 1797, he produced a globe of the Moon that he named the *Selenographia*. This sphere was fitted in most instances with a brass mechanism that allowed its users to model libration (the apparent 'wobble' of the Moon, which means that through cumulative observation more than half of the Moon's surface can be observed, and that any at one time a slightly different portion of the lunar surface is visible from Earth). As a trained artist, Russell had designed the gores himself, after developing a grid method for transferring detailed drawings of the Moon onto a spherical surface. Offered for sale, the globe could be bought either on a simple stand with no mechanism, with a mahogany ring that would allow the visible surface of the Moon at different points in its libration to be seen, or with the full complexity of the brass mechanism that Russell had developed. Intricate and costly, very few of these globes were ever produced, and no more than eleven are known to survive.

Selenographia by John Russell (1797)

In the seventeenth century, the telescope had revolutionised the way in which the Moon was understood. In the nineteenth century, some observers thought that photography might do the same. Photographic techniques were developed in the first half of the century, and were quickly put to astronomical use. While lunar photography became an art in its own right, for selenography it was important because of how it could be incorporated into the practices of map making: allowing for major features to be placed in relation to each other with greater ease, giving an overall picture of general detail. In the 1860s, suggestions for producing more detailed lunar maps included recommending the use of photographs 'as *a basis of form* on which to construct eye draughts'. They also emphasised that using such a form, 'eye-drawing ... will produce results as to details of the moon's peculiarities which light-pictures alone can never reach, because of the complexity of interpreting observations at the telescope'.[5]

Indeed, most observers were clear that photography had limits when it came to studying the Moon, and as photography developed, so too did clearer statements about how and where it was useful to the art of selenography. Unlike photographing nebulae, or faint stars, where large telescopes and photographic technologies enabled more light to be captured through the use of large lenses than the human eye could perceive, many felt that lunar detail was still best observed by eye. William H. Pickering (1858–1938), one of the most prominent lunar observers of the beginning of the twentieth century, suggested that 'as regards that which is really most interesting upon the Moon – the finer detail, and more delicate features – the photograph does not even hint at their existence'.[6]

In the 1870s, however, a book featuring photographic prints full of delicate features and fine detail had already been published: James Nasmyth and James Carpenter's 1874 *The Moon: Considered as a Planet, a World, and a Satellite*. One review of the book described the 'exquisite delineations' as 'beyond all praise'.[7] Another, by a distinguished astronomer, asserted that 'rarely, if ever, have equal pains been taken to ensure such truthfulness'.[8] Most remarkable was not that the images were photographs of the Moon itself, but that they were of painstakingly constructed plaster models, bringing together some 30 years of observations.

Plaster model of the lunar Apennines, Archimedes, and surrounding area by James Nasmyth

These lunar images were technical and artistic marvels, intended to describe a technical and artistic marvel of a different order. In the mind of James Nasmyth (1808–90), a Victorian engineer, most famous for his invention of the steam hammer, metal founding was both a practical and a poetic analogy for the formation of the Moon. Practical, because it was Nasmyth's knowledge of the way metal behaved as it cooled that led him to his volcanic theory of the lunar body. Poetic, because to him it was also a stable testament to the wonder of a creative God: the Moon, he wrote, was 'medal of creation,' with 'every vestige sharp and bright as when it left the Almighty Maker's hands'.[9]

The Moon, he suggested, was at one point a molten body. The surface had cooled and solidified. As its interior began to do the same, it expanded, as metal does just before the point of solidification, and immense internal pressure was generated. Cracks appeared in the lunar surface, and lava spilled out to form vast plains. Volcanoes erupted, scattering debris and streaming lava. On finally solidifying, however, the volume of the lunar interior shrank away from the crust, causing it to fracture, wrinkle and fold. Since then, in the absence of an atmosphere, the Moon had remained unchanged. Nasmyth wrote 'it arrived at its terminal condition aeons of ages ago, and ... we are presented with the sight of objects of such transcendent antiquity as to render the oldest geological features of the earth modern by comparison'.[10] Marshalling arguments about how he had reached such conclusions, including the way that the skins of apples wrinkle as the fruit shrinks, he suggested that 'Where the material eye is baffled, the clairvoyance of reason and analogy comes to its aid'.[11]

Reason and analogy had always shaped how the Moon was understood. In the seventeenth century, a strong and lasting analogy between the Earth and the Moon was developed. The first major implication of this was that once the Moon was considered Earth-like – hills, craters, evidence of past change – it lent weight to an understanding of Earth not as a unique centre of the cosmos surrounded by a perfect and unchanging celestial sphere, but as a planet among others, themselves marked by change.

To astronomers and natural philosophers, the analogy between the Earth and the Moon did not only contribute to a shift in cosmological understandings. The more observers thought the Moon to be world-like, the more they began to

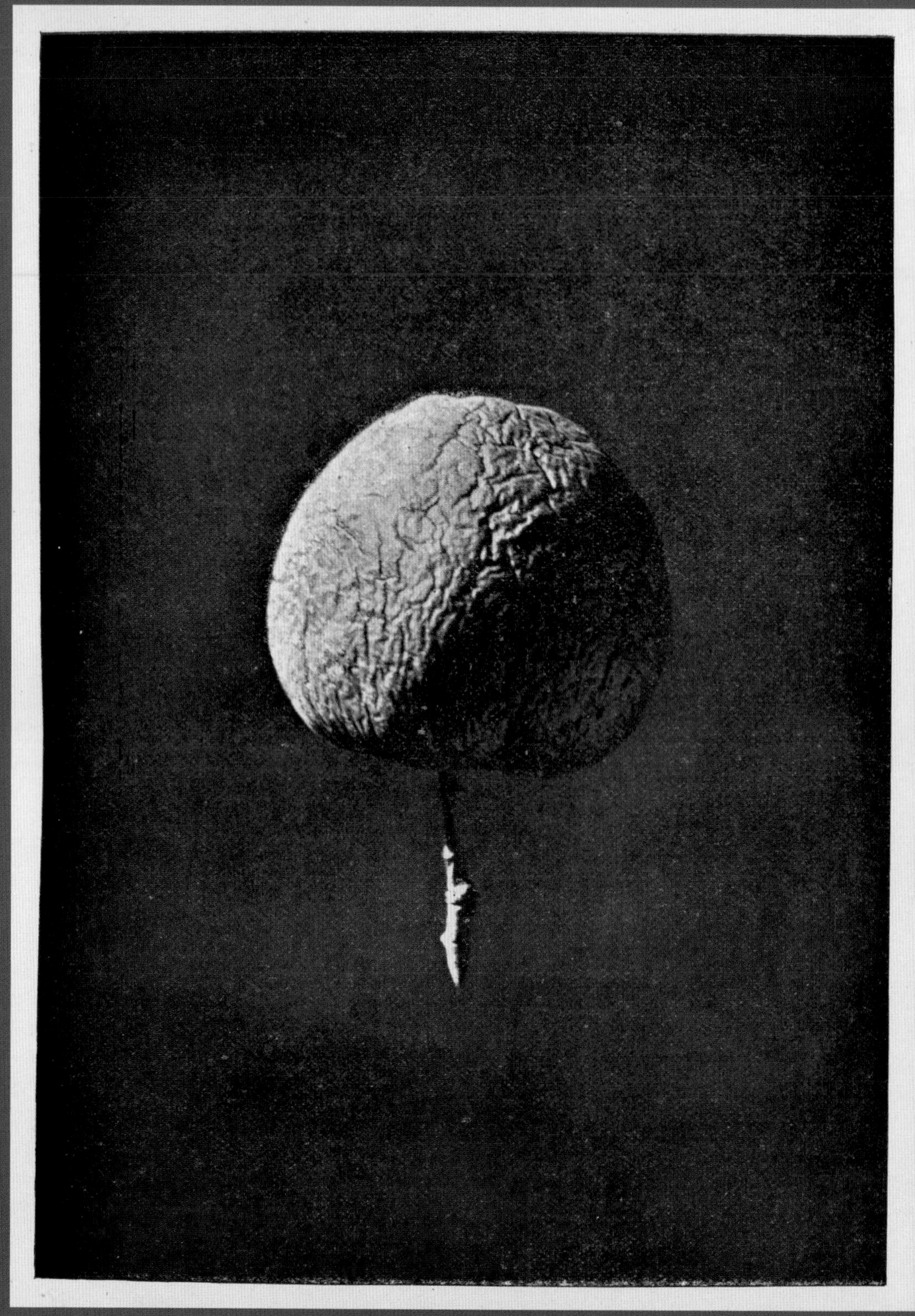

'Shrivelled apple, to illustrate the origin of certain ranges resulting from shrinkage of the interior of the globe' from James Nasmyth and James Carpenter, The Moon: Considered as a Planet, a World, and a Satellite, *4th edition (1903)*

speculate on what could be learned about its nature from telescopic observation. Adrien Auzout (1622–91), Parisian astronomer and one of the founding members of the Observatoire de Paris, mused on what terrestrial changes 'the supposed inhabitants of the Moon might discover' by looking at the Earth, 'to see, whether reciprocally I could observe any such in the Moon'. The colours of the fields in different seasons; forests cut down or marshes drained; sea ice as it formed, and as it melted: all, he suggested, changed the Earth's surface enough to be seen by hypothetical lunar astronomers. By implication, astronomers on Earth should be able to detect similar changes if such took place on the Moon. Based on this reasoning, he called for greater efforts in lunar cartography: better Moon maps would enable future observers to detect change over time, because past maps could be compared with the Moon itself.[12]

Over a century later, lunar observers reported seeing changes on the Moon in real time. William Herschel (1738–1822), musician and eminent astronomer, wrote of seeing volcanoes. These were not extinct craters, but burning details on the lunar surface, which, through his telescope, 'exactly resembled a small piece of burning charcoal,' with 'a degree of brightness, about as strong as that with which such a coal would be seen to glow in faint daylight'.[13] Elsewhere, Herschel wrote of seeing forests, with trees waving in the lunar shade, and stressed the importance of analogy in his lunar work: 'as in things beyond the reach of Observation we have no other way to come at knowledge, the imperfection of these Arguments may in some measure be excused; And I may venture to say that if we do not go so far as to conclude a perfect resemblance, we must allow great weight to inferences taken from this source.'[14]

This emphasis on analogy would only increase in the nineteenth century as more and more attention was paid to the natural history of Earth's satellite. It was a century in which geological thinking was revolutionised by the proposal that the Earth was not thousands, but millions of years old. As geology flourished, what, people wondered, could be known about the formation of the Moon? Were her features still changing? How far could knowledge about the Moon be based on knowledge of terrestrial geology? And, as for so long, the looming question that underpinned investigation of the Moon: how much can be claimed about a place where no one can go? German astronomers Beer and Mädler, who in 1836 published the most detailed lunar map up to that point, which would become a key reference work, famously claimed that the Moon was 'no copy of the Earth' and argued that 'there must be an absolute difference in every respect ... the economy of organic nature must be as different as the surfaces'.[15]

For much of the nineteenth century, the interpretation of the Moon was as a dead world, which developed into a rich seam of gothic description. John Pringle Nichol, Regius Professor of Astronomy at the University of Glasgow, wrote a popular astronomy book, *Contemplations on the Solar System* (1844), in which he took the reader on a tour of the crater Tycho. His description is heavy with portent and scenes of past destruction:

> *Ascend, then, O Traveller! Averting your eyes from the burning sun; and having gained the summit, examine the landscape beyond! Landscape! It is a type for the most horrible dream – a thing to be thought of only with a shudder. We are on the top of a circular precipice, which seems to have enclosed a space fifty-five miles in diameter from all the living world for ever and ever! ... Off then, down and arrive! It is indeed a terrible place! There are mountains in it, especially a central one 4,000 feet high, and five or six concentric ridges of nearly the same height, encircling the chasm; but the eye can rest on nothing except that impassable wall without breach ... nothing here but the scorching sun and burning sky; no rain ever refreshes it, no cloud ever shelters it...*[16]

But not all lunar features were alike, and Nichol emphasised that the sort of attention being paid to determining the timescales on which different geological features were formed on Earth could also be paid to the lunar body. Acknowledging a long history of formation, and dividing lunar features accordingly, also had potential to uncover the earliest terrestrial history, revealing forces that had long ceased to be detectable on Earth.[17] How this attention to lunar geology developed in the twentieth century is explored by David A. Rothery later in this volume.

Within the variety of lunar projects that selenography enabled, whether developing geological accounts of particular features, or constructing lunar maps to determine terrestrial meridians, naming became integral to the process of mapping. Developing a shared language to describe the Moon's features facilitated astronomical work, something essential in a form of enquiry that depended on so many different eyes, hands and minds. Van Langren, whose 1645 map was made under the patronage of Philip

'Pleniluni lumina austriarca philippica' by Michael van Langren (1645)

IV of Spain, combined the language of terrestrial features – marshes, oceans, rivers, bays, mountains – with the names of powerful figures in the Habsburg court, contemporary astronomers and a handful of saints.

Hevelius, the great Polish Lutheran astronomer, rejected the nomenclature that had come out of the heart of Catholic Europe, concerned that naming features after contemporaries would only sow unnecessary discord. What he proposed was a nomenclature system based on a particular set of terrestrial features, making the Moon a mirror of the Earth. Searching for a suitable region on which to base this lunar geography, he wrote 'I found to my perfect delight that a certain part of the terrestrial globe and the places indicated therein are very comparable with the visible face of the moon and its regions, and therefore names could be transferred from here to there with no trouble and most conveniently.'[18] Thus, on Hevelius's map one finds features named after the Mediterranean, Adriatic and Caspian Seas, the River Nile, the islands of Sicily and Sardinia, and many more. Hevelius's vast network of correspondents across Europe (to many of whom he sent copies of his *Selenographia*) helped the widespread adoption of his naming system.

When Riccioli came to develop a naming system, he was unconvinced by the correlation between lunar and terrestrial features in Hevelius's approach. Equally unwilling to name lunar features after contemporary courtiers, as van Langren had done, he developed a system using the names of figures connected with astronomy for craters and smaller features. For instance, craters are named after Azophi, the tenth-century Persian astronomer al-Sufi, and the ancient Greek scholars Eratosthanes and Aristarchus. They are also named after more modern scholars, such as Copernicus (for Nicolaus Copernicus, 1473–1543) and Fontana. Terms relating to terrestrial weather, such as the Sea of Rains and the Bay of Rainbows, were given to the

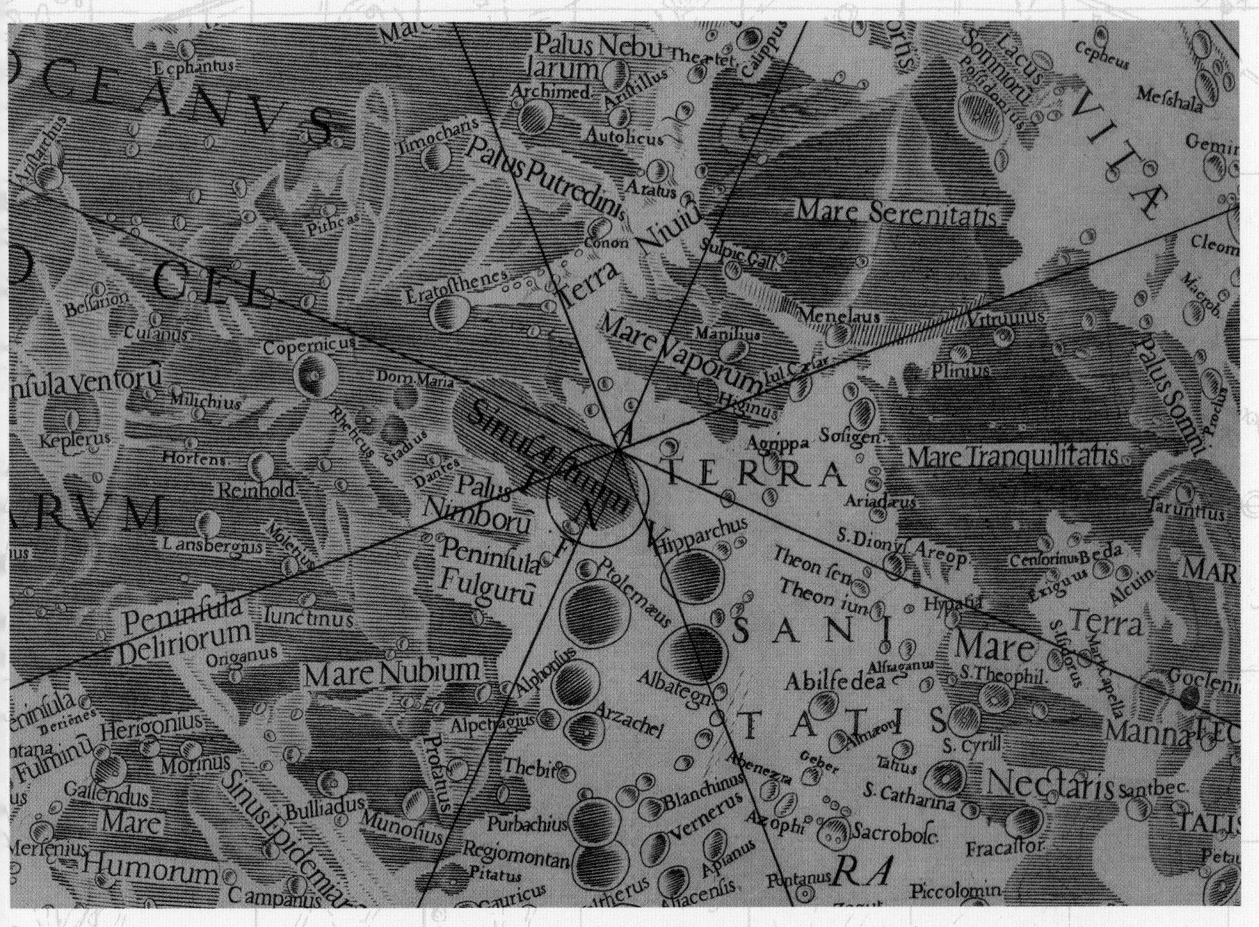

Detail of 'Figura pro nomenclatura et libratione lunari' by Francesco Grimaldi and Giovanni Riccioli, Almagestum Novum *(1660)*

larger plains, or seas. For much of the eighteenth century, communities of astronomers would use either Hevelius's or Riccioli's nomenclature. In the nineteenth century, the use of Riccioli's in a major new selenographical publication, that of Beer and Mädler, established that nomenclature as the predominant system.

As telescopes became more powerful and Moon maps more detailed, so more features needed distinguishing by name to allow observers to share their work in a meaningful way. Some numbers from the nineteenth century help make that clear: Beer and Mädler's Moon map (1836) had 7,735 craters. Another, by astronomer Julius Schmidt, published in 1878, had 32,856. Astronomers began to designate lunar features as they did clustered stars, assigning letters – Roman letters to craters, Greek letters to peaks – to help distinguish different parts of what had once been understood as one feature. Even so, the situation became confused, as the same features were assigned different names by different people, making the collaboration between observers and observatories at the heart of astronomical work increasingly difficult.

It was to this problem that astronomer and mathematician Mary Blagg (1858–1944) turned in the early twentieth century. In her mid-40s, she had attended a course on astronomy, and her astronomical, mathematical and linguistic skills were such that the course lecturer and well-connected astronomer, Joseph Hardcastle, invited her to participate in one of the great selenographic projects of the time, resolving lunar nomenclature. So impressive was her work, first published in 1913, that she was among the first four women to be elected Fellow of the Royal Astronomical Society in 1916. A further development of her work, co-authored by Austrian astronomer Karl Müller, *Named Lunar Formations*, was published in 1935. Alongside this came the official lunar map of the International Astronomical Union, also co-authored by Blagg, which became the standard reference work on lunar nomenclature for the next 30 years.[19]

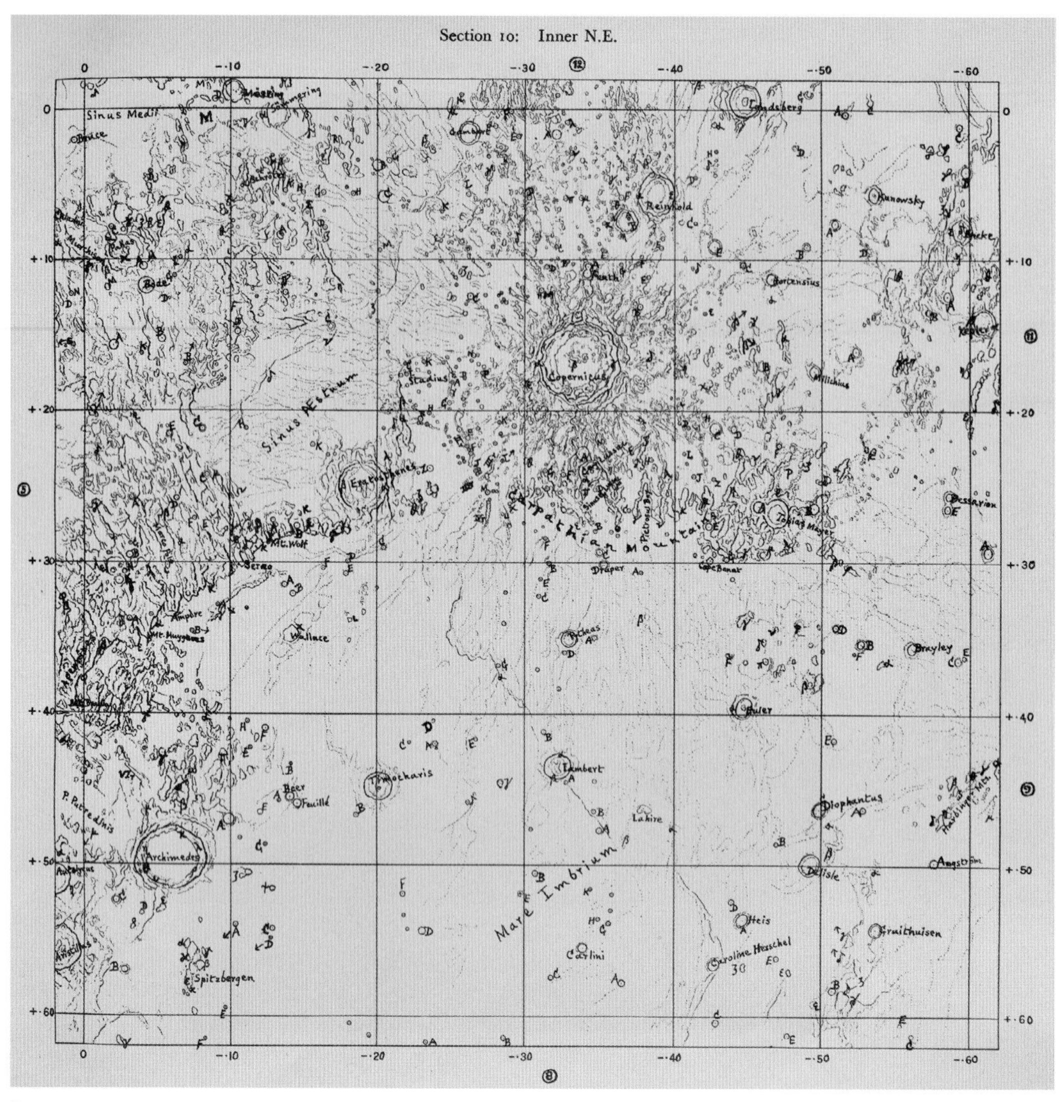

NNE sheet from Mary Blagg and Karl Müller, *Named Lunar Formations*, vol.2 (1935)

Blagg's work highlights the continuing prominence of unpaid astronomers in producing, developing and contributing to maps of the Moon. As increasing numbers of serious amateur astronomers turned their telescopes to the Moon, articles in astronomical periodicals gave hints and tips on how to produce not just accomplished, but useful lunar drawings. Observers were advised not to choose an area of the Moon that would take more than 45 minutes to an hour to draw, as a longer period would involve too much change, not only in the way that light fell on the Moon, but also in the way the atmosphere irregularly interfered with what could be seen. Hugh Percy Wilkins, president of the British Astronomical Association's Lunar Section 1946–54, and perhaps the mid-twentieth century's most important lunar observer, wrote that making a lunar map:

> *requires a critical examination at sunrise and at sunset over a long period ... It requires many sketches and drawings if it is to be of any value at all, and above all, it requires a great deal of patience and time. The observer has to watch the sun rising right form the earliest dawn, to pick up details as they may be uncovered by shadow. Much of the finest detail, the most difficult to depict, but by far the most fascinating, vanishes before one's eyes, dying, so to speak, even while it is born.*[20]

PTOLEMÆUS.

Feb. 26. 1939. 6 in. Spec. X100

Clear sky & good definition.. floor very dark, almost black.
7 objects on floor.

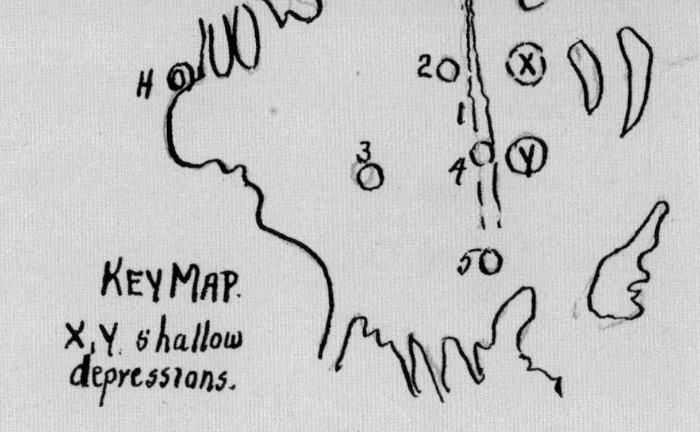

'Ptolemaeus Feb 26 1939' by Hugh Percy Wilkins

Wilkins's own Moon drawings reveal the attention that he gave to this fine detail. In thick black ink he illustrated the way in which shadows fell from a particular lunar feature over the course of time. Through this process he sought to reveal the character of different parts of the Moon more closely. As he did so, he developed ideas about change that drew on 'the new selenography' of the earlier twentieth century. Departing from the 'cold dead world' understood by most nineteenth-century observers, this insisted on 'real, living changes' on the Moon's surface.[21] In line with this, Wilkins's work placed great emphasis on observing ever finer detail, leading later critics to complain of a cluttered quality to his maps. His approach went rapidly out of fashion, as emphasis shifted to larger-scale interpretation of the Moon. Still the focus on detail meant that in the middle of the twentieth century Wilkins's maps proved useful in the interpretation of lunar photographs, made both on Earth and in space. In 1959, for example, his maps were used by Soviet astronomers to identify features from the edge of the Moon's visible surface from photographs taken from *Luna 3*.[22] In 1963, NASA ordered a copy of Wilkins's 100-inch map of the Moon, and it was incorporated into the resources used in the Apollo programme.

Within government mapping programmes, the emphasis was somewhat different. Cold War politics demanded impressive visual and interpretive firsts: images of the far side; sophisticated geological work. Perhaps the biggest Moon coup of the Fifties had been the images of the far side produced by the Soviet spacecraft *Luna 3*, on 6 October 1959. Transmitted to Earth by radio, the images gave an initial impression of a 'monotonous' surface because of the absence of large maria, and the blurriness of the photographs.[23] A triumph of both spaceflight and image making, the data produced were quickly included in an *Atlas of the Far Side of the Moon* (1960). In 1961, a globe was published by the Moscow Central Scientific Research Institute, though without coverage of the whole of the far side, part of it remained blank. Even as it was published, this globe was going out of date, as new information became available.[24] This would become a common theme in lunar cartography in the Sixties: intensifying observations from Earth and new data from Moon-oriented spaceflight produced more and more information to be incorporated into maps for scientific as well as popular use.

It was in the Space Race context too that both the USSR and the USA began producing geological maps of the Moon, using photographs from the large observatories and visual telescopic observations. As the unmanned lunar exploration programme intensified, data produced in space was incorporated into this geological mapping project – this included photography, laser altimetry (lasers

Detail from 'Section VIII' of Hugh Percy Wilkins's Map of the Moon *(1952)*

being bounced off the lunar surface to measure the height of different features) and magnetometry (measuring the Moon's magnetic field). These maps became part of the planning for landing sites in the Apollo programme, as well as being part of new interpretations of the natural history of the Moon and its relationship with the Earth.

Mapping the Moon had, over the centuries, given people new ways of reflecting on it, as well as new challenges: how to interpret and describe sets of observations made through lenses; how to discern the nature of the Moon without going there; how to communicate with other observers what had been seen; how to use the results to understand Earth's place in the universe, as well as our place on Earth. Asked in different forms over hundreds of years, most of these questions remain central to how we understand the Moon today.

Lunar table globe, 1961

Portraying the Moon: an Artist's Eye

Melanie Vandenbrouck

John Russell (1745–1806) remains the foremost English pastel artist. A prominent portraitist in his day, he produced hundreds of likenesses of the rich, famous and fashionable society members of his time.[1] A much smaller group of privately produced and more unusual 'portraits' chart his enduring obsession with the Moon. His lunar corpus includes at least seven pastels, including the largest picture of the Moon produced up to that time, extensive sketches, two lunar planispheres and his mechanical lunar globe or *Selenographia*.[2]

Although his earliest known lunar drawing dates from 1764, it was in the mid-1780s that observing the Moon became a regular night-time pursuit, one that would last for almost two decades. Russell counted leading astronomers among his friends, including the King's Astronomer, William Herschel (1738–1822) and the Astronomer Royal, Nevil Maskelyne (1732–1811). It is to the equally eminent President of the Royal Society, Sir Joseph Banks (1743–1820) that he reputedly owed his lunar pursuits, developing Banks's suggestion that the features of the Moon could only be adequately rendered by an artist.[3] In 1788, Russell exhibited at the Royal Academy a portrait of Banks spiritedly holding a sheet on which a luminous Gibbous Moon stands out, with 'Carte de la Lune par J. Russell' inscribed on the edge of the sheet. The movement, suggested by the curled paper, gives the sense that Banks is engaged in animated conversation, perhaps about Russell's own astronomical endeavours. Though Russell kept his Moon pastels to a select group of scientists in his circle, by exhibiting Banks's portrait at this most public of art institutions, he asserted the natural philosopher's admiration for his lunar work.

Russell's initial attraction to the Moon was an aesthetic one, as revealed in a 1789 letter to Observer at the Radcliffe Observatory, Oxford, Thomas Hornsby (1733–1810). Recalling first seeing the Moon through a telescope 'two Days after the first Quarter', he wrote:

> *you will conclude how much struck a young Man conversant with Light, and Shade, must be with the Moon in this state, especially, as I was not taught to expect such clearness and expression, as is to be found near and upon the indented Edge.*[4]

In a Waxing Gibbous view, the point in the lunar cycle he had described to Hornsby, he conveyed the drama of a world emerging out of a deep blue ether. By applying thick layers of white chalk, he highlighted the gleaming crater rims at the terminator. By contrast, he used the powdery effects of the pastel to their fullest in the more illuminated parts of the Moon, stumping out the brightest area with his finger.

Russell's ambition was no less than to improve lunar cartography.[5] Criticising the first Director of the Observatoire de Paris and author of a celebrated Moon map, Jean-Dominique Cassini (1675–1712) for 'too hastily conclud[ing] that the large parts upon the Moon's Face were seas', he wrote to Hornsby that 'some considerable room is left for improvement'.[6] Russell applied himself assiduously to his project. To Hornsby, he confessed that 'for several years I have lost few opportunities when the Atmosphere has exhibited the Object of my study and imitation',[7] while to the painter Joseph Farington (1747–1821), he claimed spending daily '6 Hours out of 24 calculating an average number, in experiments, in drawing or in making calculations'.[8] Sustained application was necessary for '[t]he Moon requires much attention to be well understood, being composed of so many parts, of different characters, so much similitude in each class of Forms and of such variety in the minutiae composing those Forms'.[9]

From the astronomers in his acquaintance, he sought advice on telescopes and lunar calculations. He used sophisticated telescopic equipment, a 6 foot (1.83 metres) reflector with a 6 inch (152 mm) mirror designed and lent by Herschel, and a Dollond refractor, given by his friend, the sculptor John Bacon (1740–99).[10] Russell's approach included combining the method of triangulation and the use of a micrometer to establish the distances, relative positions and proportions of lunar features with exactitude. To this, he added the study of their shape and depth through repeated observations under different light conditions. The drawings, executed at various degrees of magnification and detail, were annotated with the date, time of observation and viewing conditions, in line with contemporary astronomical technique.

John Russell, Portrait of Sir Joseph Banks, *1788*

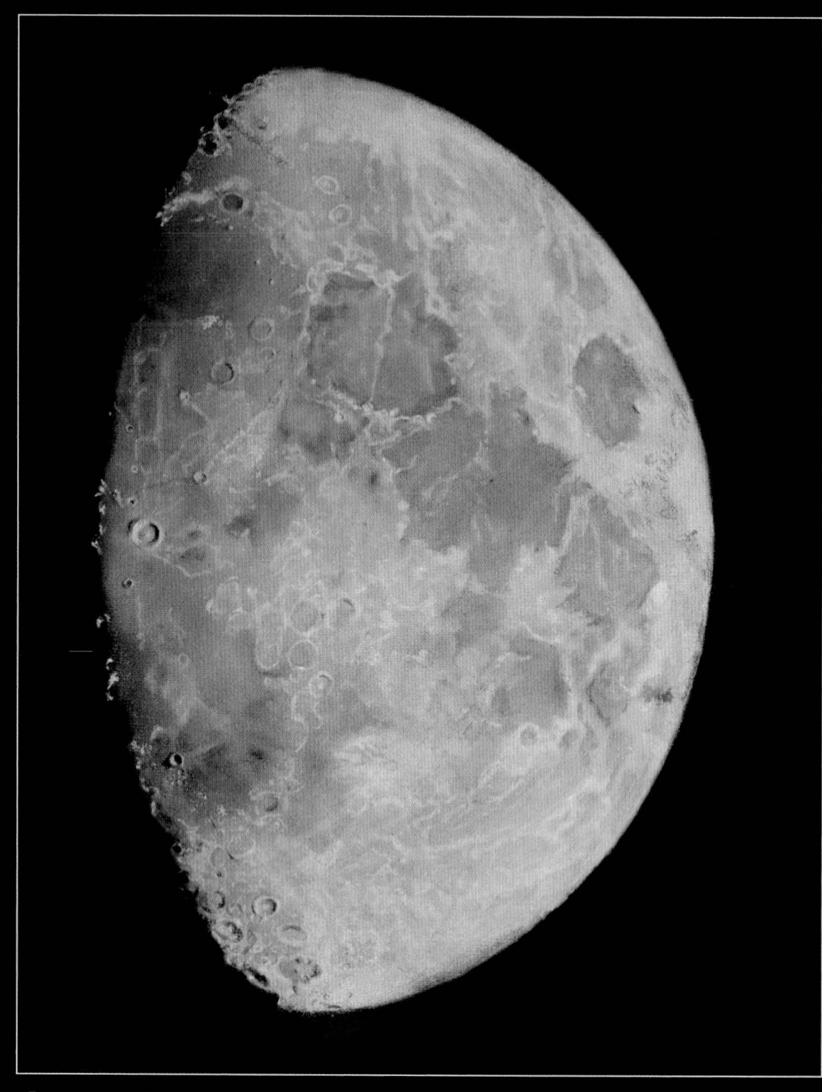

John Russell, Waxing Gibbous Moon, two days after the first quarter, c. 1787

Russell complained that, unlike the sketches executed in black lead pencil, the technique of pastel did not lend itself to accuracy. What they lacked in detail, however, they gained in feeling.[11] They recorded the 'effect' so strikingly lacking from other cartographers' attempts, in which 'the just proportion is not maintain'd in the Gradations between the inherent Light and Dark parts of the Moon, by which all pleasurable distinctness of character is produced'.[12] One pastel presents a Half Moon alongside the careful scrutiny of a conglomeration of craters, with a depth and velvety sheen that affirm the artistic quality of his scientific observations. In his treatise, *Elements of Painting with Crayons* (1772), Russell asserted that 'Painting is an art in which truth of Out-line is no less necessary than justness of Colouring.'[13] These were rules he applied in human and lunar portraits alike. Russell produced a likeness of both the physical features and the spirit of his sitters. Similarly, he endeavoured to capture the character of the Moon. He described his approach thus:

> *first ... representing the colour of the Moon in its general form according to the particular Phase, upon this I laid the larger Spots of Mediteraneum Serenitatis Tranquilitatis &c—these I adjusted by degrees giving them their general Forms and grand bearings endeavouring to preserve ... the*

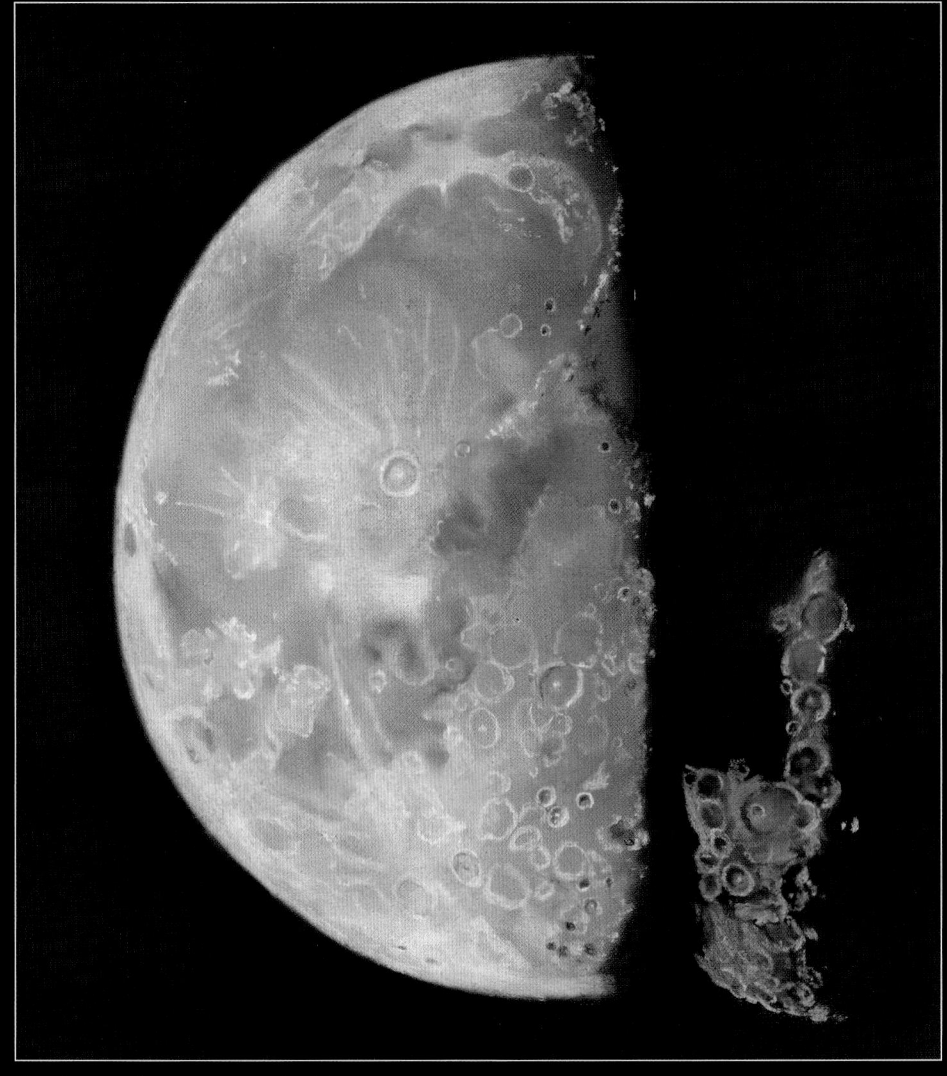

John Russell, Telescopic Study of the Moon, *1796*

proportional difference of dark & Light; ... executing the minutiæ as far as time would permit... The Effect of the Moon thus produced, surrounded by a dark Blue colour has a novelty as well as an expression in its appearance, which has given some pleasure to many [astronomers].[14]

Combining the pastel studies as a basis for 'effect', and lead drawings for detail, Russell met the challenge of obtaining an accurate rendition of the Moon by undertaking a large pastel (1795, Museum of the History of Science in Oxford), the size of which (130x150 cm) would allow for the precision lacking in the smaller depictions.[15] Although the pastel, damaged by damp, has lost the exactness Russell intended, its sense of scale and monumentality remains.

Russell favoured depicting the Moon in its gibbous state 'because at that time a greater number of eminences on the moon's surface are distinguishable than when she is in opposition to the sun'.[16] When full, its features 'very faintly express their character [...] as they are nearly lost in the general Blaze of Light'.[17] His Full Moon is thus distinctive in its relative flatness and uniformity.

Russell's ultimate ambition was to depict a view of the Full Moon with all its features distinctly legible. To this end, he exerted both artistic and astronomical license, as revealed

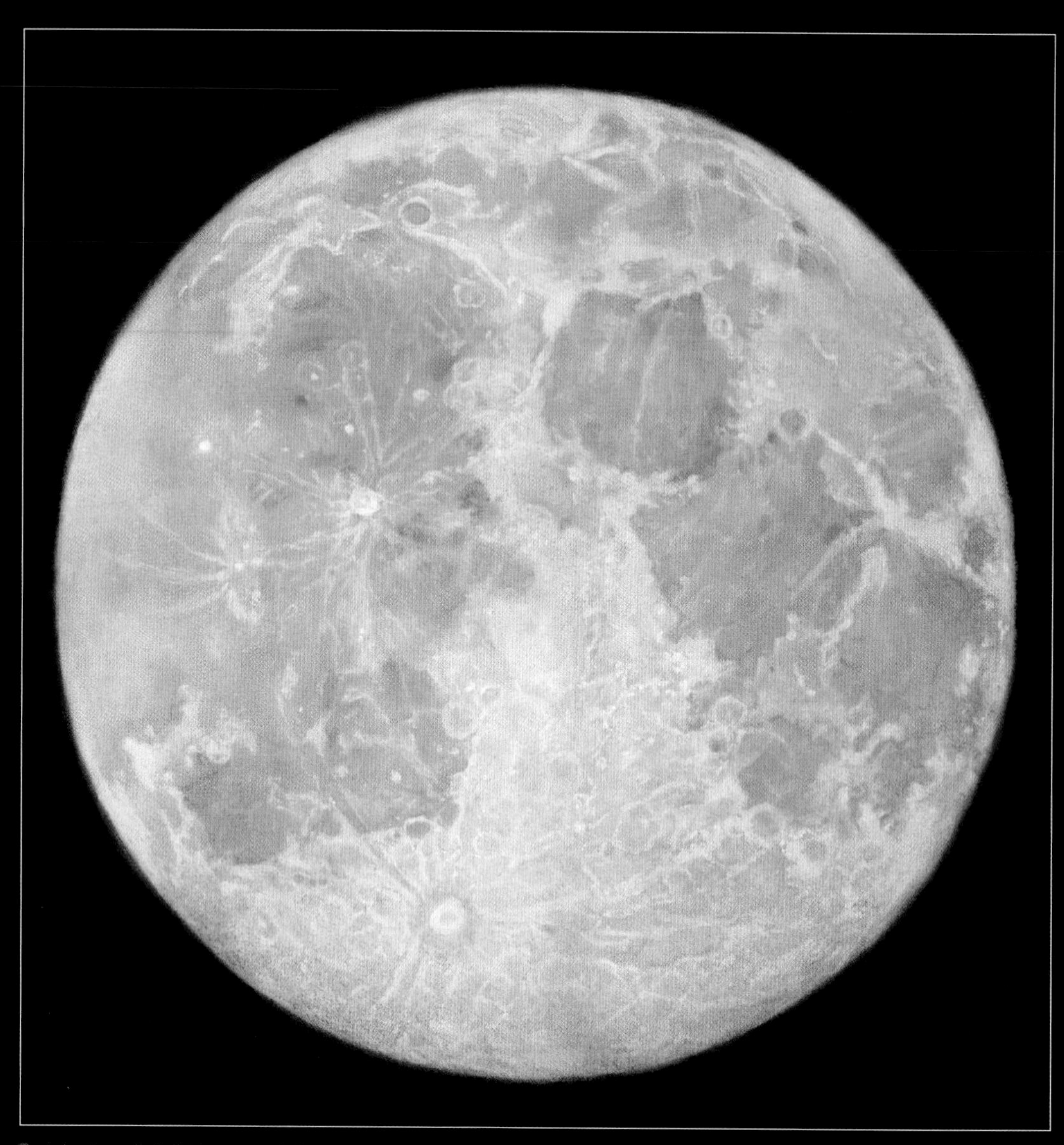

John Russell, Full Moon, *1790s*

by the two lunar planispheres published posthumously by his son in 1809. The first is a traditional depiction, the second, 'synthetised from various angles of lunar illumination', a composite of sharply defined individual lunar features when observed in oblique light. The aim of this impossible view was, Russell's son wrote, 'ascertaining the longitude of places by the transit of the earth's shadow, when the Moon is eclipsed', an approach traditionally hindered because when its face is fully illuminated, particular features lack the definition required.[18] These did not, despite his son's best hopes, 'prove of great utility to the Astronomer [and] lead to very important speculations in Natural Philosophy'.[19] Worse even, this approach went against Russell's own precept that 'the grand object [of a painter's] pursuit is … a just imitation of nature'.[20]

Russell's lunar observations had climaxed in his *Selenographia* and yet, despite Farington's claim that 'Dr. Herschell [sic] has examined it 2 Hours, and said Astronomers could not now do without it', it failed to attract interest.[21] Likewise, art historian Antje Matthews argues that because of their absence of grid and legend, the planispheres and pastels would have been of limited use, and that 'the moon images can only be regarded as knick-knacks for the gentleman collector of curiosities'.

John Russell, Lunar Planisphere, *in oblique light, published by William Faden, 1806*

hese were 'neither maps nor works of art', they ritical fortune.[22] To some, Russell's 'selenomania' ted his portrait production. One observer at the ademy exhibition of 1795 thus remarked: 'This n follows his studies of the Moon, very much advantage of his earthly pursuits'. One of his the mordant critic added, 'is a miserable daub: ng is bad; and the pencilling poor and feeble'.[23] l suffer from the pursuit he 'only esteem[ed] as ement'?[24] On the flank of an unmounted version

> *On John Russell, painter of the Moon*
> *Actaeon and Endymion did not see*
> *[the moon] closely.*
> *Them alone we remember; But where is Russell*

As Maskelyne remarked, we are more fam these mythical figures than we are with the artist, w emerging out of the Moon's shadows. Precisely they eschew categorisation, Russell's lunar portra be reconsidered as all the more valuable for their

By the Light of the Moon

Kelley Wilder

'How do you photograph the moon? Do you use the magnesium light?' Manchester photographer Alfred Brothers (1826–1912) found it absurd that he was so often asked this question, even in 1883. Photography had by then been employed for years in astronomical observing, but as Brothers was so frequently made to realise, many people found it astonishing that a photograph of the Moon could be made using only its light.[1] Photography was a difficult technology, and although it became an essential part of astronomical work in the twentieth century, the development of its early promise took years of trial and error. Indeed, photographs of the Moon had an important role even at photography's very beginnings, but achieving them took hard work, precise chemistry and unlimited patience.

Photography was first announced in public on 7 January 1839, to the assembled members of the Académie des Sciences in Paris, though by that time had existed in different forms for several years. One of the most remarkable details of this announcement was that Louis Jacques Mandé Daguerre (1787–1851), inventor of the daguerreotype being presented that day, was also the first person to have photographed by the light of the Moon. Maybe the emphasis was unsurprising, given that two of France's most prominent astronomers, François Arago (1786–1853) and Jean-Baptiste Biot (1774–1862), were in command of the introduction of the daguerreotype to the Academy members.[2] In making the announcement, Arago was making a concerted political effort to convince the government to buy the invention for the good of all science, freeing it from a restrictive patent.[3] In order to do this, he had to accomplish two things. The first was to describe the daguerreotype and its potential for the sciences. The second was to prove that it was no hoax. Daguerre, you see, was already an acclaimed master of illusion, a painter and the co-owner of the famed Paris Diorama, where in a single afternoon, from the comfort of your theatre seat, you could experience the Alps, watch the moon rise, or take in a midnight mass. Readers of the newspapers might have been forgiven for displaying skepticism at the fantastical notion of 'invisible' rays painting pictures in minutes, which were supposedly more perfect than any miniature. Such hoaxes were not altogether uncommon, and discussion of the Great Moon Hoax of 1835 (see page 146) still lingered in Parisian popular culture. The daguerreotype sounded too good to be true – but it wasn't a hoax.

The daguerreotype consisted of a finely polished silver plate, backed with copper for strength, with, at that time, a silver iodide imaging substance. After polishing, the silver plate was held over fumes from iodine crystals until it reached a particular golden colour, indicating its optimum sensitivity, before it could be exposed to light within a camera.[4] In 1839, it was said that a good exposure at midday was about 8 to 10 minutes. Of the first daguerreotype specimens on display, there was not a daguerreotype of the Moon, but this was not for lack of trying. Arago recounted that he had asked Daguerre to conduct an experiment with his plates, exposing them by the light of the Moon. Predictably, this required a much longer exposure, one far too long to secure an image of something that moved across the sky. Describing the daguerreotype process to the Academy, Arago reported that recording the Moon's rays was in itself a triumph – picturing the lunar body did not come into it. Clearly, there was a double narrative to 'photographing' the Moon.

Biot and Arago, then, were not only interested in making an accurate and mechanical (and supposedly more objective) picture of the Moon's surface. They were also interested in what sort of rays might be emanating from our satellite, which could be 'captured' on a daguerreotype plate. They wanted to test how well this new recording device would register new and unknown aspects of light. Photography had arrived in the midst of a nineteenth-century debate about the nature of light, and remained mixed up in that scientific dialogue throughout the century. It was clear by 1839 that several types of 'visible' and 'invisible' rays existed. Men and women of science quickly realised that photography, whether the daguerreotype, or later processes, were sensitive not just to the visible spectrum, but in varying ways to other rays. The daguerreotype, which was sensitive only to the blue, violet and ultraviolet end of the spectrum, promised to be the key to isolating and researching the so-called invisible rays, and thus any light

J.A. Whipple and G.P. Bond, Daguerreotype, c.1851–2

emanating from any celestial body, including the Moon. While photography became a way of detecting radiation emitted from any object, the Moon became a test object to show the difference between one photographic material and another, which was important as photographers developed and compared new recipes and processes. Photography by the light of the Moon retained this mixture of purposes until standardised photographic materials became reliable and widespread, and spectroscopy emerged as its own field. Throughout the nineteenth century and into the twentieth, the photographic image of the Moon was established as the standard, the calibrator, by which subsequent astronomical photography would be judged.

Paris wasn't the only place where scientists were trying to register images using the Moon's light. In England, William Henry Fox Talbot (1800–77) and astronomer Sir John Herschel (1792–1871) also tested Talbot's new photogenic drawing paper (salted paper brushed with silver nitrate, invented in parallel to the daguerreotype) with the Moon's rays. For Herschel, photography appeared to be a '...beautiful mode of ... mapping the moon',[5] but it would be some time before photography delivered on this promise. The bright and close Moon, with its well-charted movements, formed an integral part of amateur observing in families such as Talbot's. But creating an accurate picture was at first far beyond the reach of his chemistry, which, in 1839, could only obtain an impression of the lunar rays after 10 minutes. The Moon was a target that enabled direct comparison of Daguerre's and Talbot's processes, invented simultaneously in two countries, but which differed greatly in their method.[6] Herschel, too, had to forego his dream of mapping in favour of testing photographic materials and reported creating dark spots by the light of the Full Moon: 'Last full moon I got in 55 minutes, by light condensed by an aplanatic double lens 3in aperture & 6in focus – a very dark spot almost black. I used your Salted paper I think it would have required 5 or 6 seconds of Sunshine to have produced an equal darkness.'[7]

When Talbot announced his much-improved process, the Calotype, in 1841, in which a latent image from a short exposure was developed out with gallic acid, he was so convinced of the value of the Moon test that he dared other photographers, even daguerreotypists, most notably famed London photographer and scientist Antoine Claudet (1797–1867), to match it. 'It is a nice point to determine which is the most sensitive to light, my Calotype paper, or the Daguerreotype improved by Claudet's process.' Assuming the identity of his paper, he continued, 'I am sensitive to simple moonlight; he has not yet made the trial, which I throw out as a challenge to all photographers of the present day: viz. that I grow dark in moonlight before they do.'[8]

In New York, John William Draper (1811–82), Professor at New York University, was also hard at work using the daguerreotype process. Draper had been investigating the properties of light, when he heard about Daguerre's invention, and absorbed it into his experiments, also using the Moon as a test. Draper has long been cited as the first American photographer to make successful daguerreotype portraits of both people and of the Moon.

Few daguerreotypes of the Moon were made, and even fewer survive. Of these, perhaps none have been as widely circulated as the images made in 1851 and 1852 by John Adams Whipple (1822–91) and George Phillips Bond (1825–65) at the Harvard Observatory in Boston. Whipple and his partner James Black were well-connected commercial photographers in Boston whose close ties to Harvard University were both scientifically productive and financially lucrative. By 1851, several photographic processes were available, but Whipple was an expert with the much-improved daguerreotype and used it to attempt to capture the Moon. The original daguerreotype recipe that Arago described in 1839 had undergone significant transformation. Not only were accelerating formulae applied involving liquid sensitisers such as bromine water, chloride of iodine, bromide of iodine and others, the practice of sensitising the silver plate first with iodine, then with bromine, then again with iodine produced plates that were fast enough to record the Moon in a sharp image, using a series of exposures aligned to a clockwork mechanism.[9] Daguerreotyping the Moon was a disruptive process for a working observatory. Bond's notes of Whipple's daguerreotyping activities on 18 December 1849 and 17 July 1850 recount that the micrometer needed removing before each session. In the case of the first attempt, it was the first time the micrometer had been removed in more than a year, and consequently all observations made that evening needed adjustment.[10] Clearly, work like this required the permission and assistance of astronomers as well as the chemical and technical expertise of the photographers.

With all this difficulty, it was no wonder that although the public had heard about Moon photographs since 1839, very few people had seen one until the Great Exhibition of 1851 in London. Photographs were rare and expensive objects and astronomical ones even more so. It made them an obvious choice for display at the Exhibition, conceived as it was to promote 'the works of industry of all nations'. Whipple's daguerreotypes, when new and untouched by time, would have depicted the Moon hanging in a perfectly polished black sky. Daguerreotypes have astonishing resolving power, and even examined under magnification, they show minute detail thanks to the tiny and rounded image-forming particles that make up the image. Indeed, it was the daguerreotype's ability to resolve detail that made it attractive for astronomical work. The small, quarter-plate size of these daguerreotypes also made them more jewel-like, and more surprising. Upon opening the leather case in which it was enclosed, the daguerreotype would reveal a picture of the Moon that appeared sharper and more fixed than the real thing seen by the naked eye through the atmosphere, all packaged up to fit in the palm of your hand.

Maps, models and images of the Moon were popular exhibits at the Exhibition, and Whipple's daguerreotypes were only one example of images of the Moon on show. Charlotte Readhouse (dates unknown) of Newark-on-Trent exhibited a detailed globe of the Moon, Henry Blunt (1806–53) showed an electrotype copy of a plaster model of part of the lunar surface, and engineer James Nasmyth (1808–90) exhibited a six-foot-diameter painting of the Moon's surface.[11] Whipple was not even the only daguerreotypist to show an image of the Moon. Claudet, whose popular portrait studio off the Strand was known to many Londoners, produced a much remarked-upon exhibit of photography including daguerreotypes of solar spectra, the Sun, and the Moon on a clear night.[12]

It is all the more remarkable that multiple Moon photographs were shown in 1851 when you consider the scarcity of such objects. To date we know of only five or perhaps six individuals who attempted daguerreotyping the Moon, including Daguerre.[13] They were, consequently, in great demand, which was problematic because daguerreotypes were always singular objects, unlike photographic prints that might exist in multiple copies made from the same negative. The only way to copy a daguerreotype was, and is, to make a copy photograph. It was in this context that Whipple managed to monetise the fame of his Moon images with a process of making a glass negative from the daguerreotype, and from it a salted paper print. He called it the crystalotype, patenting it in 1850.

John Werge, daguerreotypist and innovator, managed to acquire three of Whipple's crystalotypes of the Moon, to show again at the British Association for the Advancement of Science meeting at Glasgow in 1855. Despite their rarity, then, these very few daguerreotypes had a lasting effect on the public's understanding of photography of the Moon, and lent impetus to projects attempting a full photographic mapping of the lunar surface. It was clear that the circulation of the images was critical for astronomers. It was also clear that in the burgeoning market for visual images of all sorts, there was money to be made.

In a different part of the Great Exhibition, soon-to-be astrophotographer Warren De La Rue (1815–89) was exhibiting his newly-invented machine for folding envelopes at his company stand. De La Rue claimed that seeing Whipple and Bond's daguerreotypes of the Moon inspired him to try his hand.[14] De La Rue was neither a professional astronomer nor a professional photographer, but he had the means and the technical knowledge to do both. He took up photography in the 1850s, using the newest technology available – wet collodion on glass and later the so-called dry collodion technique. Collodion on glass refers to the process used for making the negative, and these negatives would then be printed, most frequently on albumen paper, but sometimes also on salted paper, or back onto glass, for projecting as lantern slides, or reproducing in print. The collodio-bromide emulsions used were very fine grained, so they worked well for enlarging. The photographer cleaned and polished the glass (or ran the risk of having the image peel off it later), flowed the collodion onto it, tipping it slightly from one side to the next to ensure good coverage, then dipped it into the silver solution, inserted it in the holder and made the photograph. After exposure, the plate needed developing, washing, fixing, more washing and drying before it could be assessed as successful or not (and if it was not successful, the used emulsion could be peeled off the glass and the glass re-used for another exposure). By the 1880s, using this process meant that a Full Moon on a winter's night could be recorded in about one second. Working at his observatory in Cranford, Middlesex, using a 13-inch reflecting telescope, De La Rue managed to secure numerous negatives of the Full Moon, and various phases, between the years 1858 and 1862.

Like Whipple, Warren De La Rue found a way to circulate his photographs far and wide, as part of the newest photographic craze, the stereograph. The use of stereo devices with drawings predated photography, but almost as soon as photographs could be made, people tried to make stereo images from them. The stereo photograph, or stereograph, is made of two separate images of the same object, taken a short distance apart, and printed side-by-side. The viewer looks at this through a device called a stereoscope, through which both images resolve themselves into one view, giving the appearance of being in three dimensions. To create stereographs, all types of photographs could, and were, used, including daguerreotypes, albumen or salt prints on paper, and collodion positives on glass. But the Moon posed a problem. To make effective stereographs of the Moon, it was not possible to make two exposures at the same moment side-by-side, as was the usual practice for landscapes or genre scenes. The astrophotographer often had to make a first photograph, and wait to make the second photograph until the Moon took up exactly the correct position again, possibly years later. As a consequence, many lunar stereographs have Full Moon pictures from different years, combined later by the photographer or publisher to ensure a good view. The stereo card on the right shows a left-hand image made at 10.12 p.m. on 27 August 1860, and a right-hand image made at 8.30 p.m. on 5 December 1859. The instrument makers Smith, Beck & Beck would enlarge the originals into carefully aligned prints to make a seamless 3D image. Within the scientific community there was a willingness to exchange images for the purpose of promoting scientific knowledge, but the driver for making these images, in the case of stereographs, was commerce. Photographers consequently sought to profit by registering their valuable images for copyright.

Between 1862 and 1914, photographers or agents who had paid a photographer for his or her negatives could claim a copyright on their photographs by registering them at Stationers' Hall. On 15 October 1862, Warren De La Rue took out his first photographic patent for the left- and right-hand images of a stereograph of the Moon. Selling the copyright to Joseph Beck, a prominent publisher, allowed De La Rue the freedom to make and license his own Moon photographs for broad distribution, as publishers sold photographs through their own booklists. De La Rue would take out a further three patents, each for two distinct Moon images, over the following year and a half.[15] Eventually, Smith, Beck & Beck published a series of six stereo cards of the Moon by De La Rue, gathered together in an attractive wallet. They were not the only photographer/publisher combination to produce stereos of the Moon, and indeed the legacy of copyright can tell us quite a lot about photographic business in the later nineteenth century.[16] For instance, the stereograph below made with Henry Draper's negatives of the Moon was published by Bierstadt of Niagara Falls, but distributed through one of the largest stereo card distributers in the world, Underwood and Underwood, some of whose

Henry Draper, Stereograph of the Moon, *published by Charles Bierstadt*

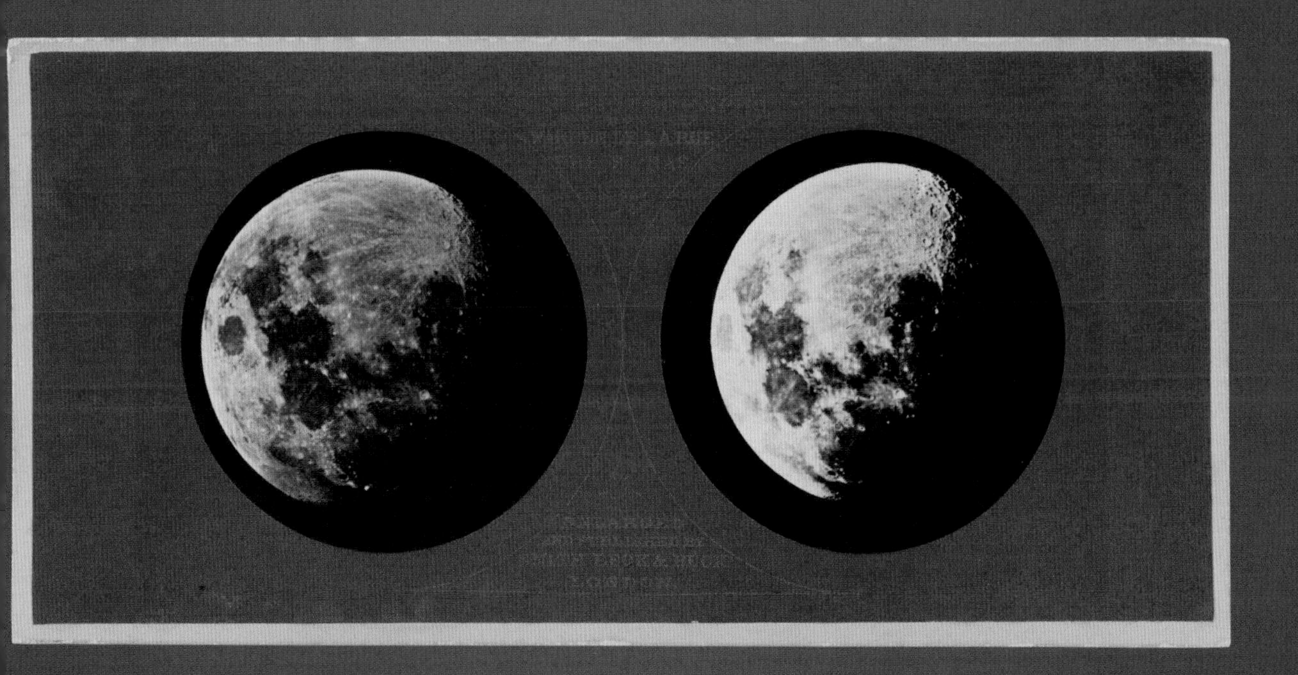

STEREOSCOPIC SERIES.—N°. III.

LEFT-HAND PICTURE.	RIGHT-HAND PICTURE.
DATE OF NEGATIVE, 1860, Aug. 27, 10^{h} 12^{m}.	DATE OF NEGATIVE, 1859, Dec. 5, 8^{h} 30^{m}.
AGE OF MOON, 11 days.	AGE OF MOON, 11·3 days.
LIBRATION IN LONGITUDE .... $+5^{\circ}$ 14′	LIBRATION IN LONGITUDE .... -5° 48′
LIBRATION IN LATITUDE ...... $+1^{\circ}$ 24′	LIBRATION IN LATITUDE ...... -5° 29′

[ENTERED AT STATIONERS' HALL.]

Warren De La Rue, Stereograph of the Moon published by Smith, Beck & Beck (recto and verso)

Photograph of James Nasmyth's plaster models and a telescope, 1858

stock was later absorbed by an even larger company, Keystone View, which marketed Moon stereographs to schools and universities across the world until it folded.

But the Moon, photographed, was not only a novelty, commodity or educational tool. It was a promise of scientific investigation of a new sort, using a new and rapidly changing technology. As photography changed, so did depictions of the Moon. In 1874, a colleague of De La Rue's, the engineer James Nasmyth, published one of the first books on the Moon to use photographs as illustrations – *The Moon: Considered as a Planet, a World, and a Satellite*. The well-known photographs scattered throughout the book are evidence of how difficult it was to research something so three-dimensional through the still-developing medium of photography.[17] Instead of photographing the face of the Moon through a telescope, Nasmyth made plaster models of lunar craters, and then had them photographed in raking light to show the distinctive shadows by which many craters were known.

He did this for two very practical reasons. Photographers were still restricted by a photographic technology that was only partly sensitive to the light of the Moon, and tricky to handle. Each collodion wet-plate negative had to be coated by hand just before exposure. Although dry collodion, using humectants like honey, beer and albumen to keep the emulsion moist, could be batch prepared for astronomical events, it was not practical.[18] Cost was an even greater hindrance. Astrophotography was an extremely expensive pastime, reserved for the landed classes who possessed the required time and resources. Talented and privileged amateurs like De La Rue, or the Welsh family of photographers and scientists, the Dillwyn Llewelyns, had access to their own observatories at their homes, but such activities were not typical.[19]

The complicated contortions needed to produce photographs of the Moon were partly determined by wet plate technology. Success in the complicated technical details was largely dependent on the skill of the operator, and many of the processes were too tricky to capture photographs of the Moon directly.

What was really needed was a way of pre-preparing plates that could be manufactured and shipped, saved and measured, and above all relied upon to produce comparable photographs of the Moon's surface. Two new developments, one technological, one methodological, allowed for just such reliable images. In 1871, just as photographs of Nasmyth's crater models were going to publication, physician and photographer Richard Leach Maddox (1816–1902) published a method for replacing the collodio-bromide emulsion on glass with a gelatino bromide on glass.[20] Because gelatin, unlike collodion, could be dried completely and re-wet, it allowed for the possibility of commercially made photographic plates that could be stored and easily transported. Like many new innovations, the original process was initially of debatable use, but within a decade commercially produced gelatin dry plate negatives revolutionised astronomical photography. It did this in two ways. First, companies such as Eastman's Dry Plate Company (later Kodak), Paget, and the Lumière Brothers began to produce plates in standard sizes, in batches that were mostly consistent in their sensitivity, and were eventually subject to standardisation. Second, companies and individuals began to experiment on the plates by adding substances or heating or cooling them, to render them more sensitive. In particular, the heating process during the manufacture of the plates, called 'ripening', altered the sensitivity considerably. Although each photograph was still a 'silver gelatin' negative, each company experimented with different formulae to produce rapid and extra-rapid plates, many of which were tested by astronomers. Collaboration between astrophotographers and photographic research laboratories such as the one George Eastman founded at Kodak in Rochester, NY in 1912 loomed on the horizon. But the move to gelatin was only one of the innovations important to late-nineteenth century photography of the Moon.

Astronomers were also altering their instruments to make better observations, and more importantly, to make better photographs. Amateur astrophotographer Andrew Ainslie Common (1841–1903) improved his telescopes until he was able, in 1881, to photograph the fast-moving object Comet b with his 36-inch reflector. With the same telescope, he worked on photographing the most difficult target in the sky, the Orion Nebula, using long exposures. In 1883, he succeeded with a photograph that was to secure his reputation as an astronomer and astrophotographer. The nebula photograph and his subsequent photographs of Saturn, Jupiter and the Moon were exhibited beyond the confines of astronomy, at the 1883 and then the 1884 Annual Exhibitions of the Royal Photographic Society of Great Britain, and won him the Gold Medal from the Royal Astronomical Society.[21] It was Common's belief that these photographs were the first steps to photography overtaking, and eventually replacing, observation with the human eye.

Andrew Ainslie Common, 'The Moon', 20 January 1880

By the 1880s, photography had not yet become the sort of scientific tool for photographing the Moon that had been hoped for in 1839. Regardless of the technological improvements that made them more common, photographs of the Moon still emulated what observers saw in the telescope, and had yet to become a new way of investigating our satellite where photographic method changed the rules. One project that began the process of bringing photography to the centre of lunar investigation studies was the *Atlas photographique de la Lune* made by French astronomers Maurice Loewy (1833–1907) and Pierre-Henri Puiseux (1855–1928). The project could not have differed more from previous lunar photography, where a single image (or two in stereo) encompassed the whole view of the moon as seen by a single human observer. Loewy and Puiseux set out to make a comprehensive photographic atlas that acknowledged the way photography depicted the Moon, and put it at the heart of learning and studying it.

The *Atlas*, while similar to other pictorial atlases popular at the time, was also a long-term campaign. Announced in 1894, it was eventually completed in 1910, after Loewy's death, in a series of 13 fascicles, one containing only text and the others each with five to seven heliogravures of enlarged portions of the Moon.[22] The work was meant to teach people not only how to look at the Moon and recognise its various features, but also how to think more carefully about photographs of the Moon. To do this, Loewy and Puiseux made and published multiple photographs of the same portion of the Moon, and left in many of the artefacts, like dust and scratches, from the photographic process.[23]

The slight differences between exposures of the same portions of the Moon, and the clear lack of retouching, were visual clues that were meant to instill confidence in the photographs.[24] Readers were encouraged to compare similar images to develop knowledge of the Moon's surface, not because the photographs gave unmediated access to the surface of the Moon, but because in acknowledging the limitations of looking by photograph, readers were shown how to analyse through the vision of photography. For 50 years, Loewy and Puiseux's *Atlas* remained an important text, referred to by other photographers and astronomers when developing their own projects. In many ways the *Atlas* represents the heyday of analogue photography in the study of the Moon.

The argument (although it was much contested) that photography was capable of seeing and representing the Moon differently, and better, than human observers set the stage for a new relationship between photography and observing the Moon. In the 1880s the chemist Herman Wilhelm Vogel (1834–98) had published a series of papers on the introduction of aniline dyes into gelatin plates to alter the photograph's sensitivity to light of different wavelengths. The dye acted like a filter, and soon companies were manufacturing photographic plates that were more sensitive to the tricky green and yellow areas of the spectrum. Stretching the photographic emulsion first to make it respond orthochromatically (i.e. to all visible light except red) and then panchromatically (i.e. to the full spectrum of light) eventually allowed for very specific tailoring of photographs for astrophotography. But it was not only the specifications of the emulsion that changed Moon photography forever.

Glass plates dominated astrophotography until the Seventies. Although flexible film was introduced almost a hundred years earlier into cameras (most notably by George Eastman with his first Kodak camera in 1888) the atlases and maps like Loewy and Puiseux's, or the famous *Carte du Ciel* map of the stars (project launched 1887), depended on reliable measurements. Glass is more dimensionally stable, and therefore was more reliable than the available flexible films. But glass is also heavy and breakable and could only be used from land. As dreams of photographing the Moon from outside the Earth's atmosphere grew, flexible film appeared to provide the answer. It was light and could be bent through a number of tightly packed systems while it was being exposed, developed, fixed, dried, cut and scanned. Satellite photography of the Moon, though, needed to move even further away from older and wetter analogue processes. Normally, a film is developed in liquid developer and fixed in liquid fixer, both too heavy and temperature-sensitive to send into space. For the first *Lunar Orbiter*, sent into orbit in 1966, Kodak invented a 'dry' film called 'Bimat' that could be developed with virtually dry chemistry. The lightweight system enabled photography in satellites that mapped landing sites for the Apollo missions. The films were scanned and radioed back to Earth, then reconstructed as photographs in long horizontal strips. When manned missions landed on the Moon, astronauts were also equipped with cameras and film, specially designed to withstand the rigours of space travel. In this way, analogue photography mixed slowly with digital solutions, providing hybrid mechanisms for Moon images that would form the basis for digital imaging to come. Analogue photography thus laid the foundations of Moon imaging in the twentieth century, dominating our understanding of what it meant to photograph by the light of the Moon.

Blancanus – Tycho – Schiller. Plate XLIV (1899) from Loewy and Puiseux, Atlas Photographique de la Lune

Rayonnement de Tycho – Phase Croissante. Plate XXXVI (1899) from Loewy and Puiseux, Atlas Photographique de la Lune

The Moon through the Eyes of the Apollo Astronauts

Jennifer Levasseur

Former astronaut and Apollo 17 Commander Gene Cernan (1934–2017) said of the iconic image of Earth his crew took during the mission, the last to touch the Moon: '[W]e've only seen ourselves through the paintings of artists, words of poets or through the minds of philosophers. Now we've been out there, we can see ourselves.'[1] Cernan, the last person to walk on the Moon, made clear the profound impact of astronaut photography on our conceptions of both Earth and humanity. He was hardly alone among astronauts, many of whom communicated a sense of obligation and honour in providing people around the world with these portals into the spaceflight experience. The Apollo programme yielded nearly 20,000 still images, which serve as a valuable visual and scientific record of the programme, in addition to the hundreds of hours of movie footage and television broadcasts. For those who took the photographs, they are images infused with memories and personal identity. It is through astronaut eyes that we came to see ourselves anew: from an orbital perspective, as creatures caught between Moon and Earth, and as explorers of what Apollo 11 lunar module Pilot Buzz Aldrin (born 1930) called 'magnificent desolation'.

From the iconic to the mundane, these widely reproduced photographs constitute the primary source of the collective memory of early human spaceflight, especially because they evoke familiar visual themes present throughout the history of exploration photography. Photographers and expedition leaders, guided by intentions of economic and political expansion by colonial powers in the late nineteenth and early twentieth centuries, provided visual evidence that fit rhetorical need. The byproduct of this process was thematic visual categories, often repeated through exploration of other environments on Earth and in space. With such heroic figures providing visual documentation of their experiences, the images were universally admired as symbols of great achievement during challenging times, when belief in American exceptionalism was tested at home and abroad. Still photographs, as opposed to television or movies, were accessible technical records as well as rhetorical vehicles: documents for scientists and engineers involved in making the Moon landings a reality, and symbols within the context of the history of exploration photography and the Cold War.

Apollo photographs, with such similar visual character to those of terrestrial exploration, tell viewers little of how astronauts actually captured images. Astronauts were, in every sense of the word, amateurs in the art and craft of photography, trained as they were for so many other activities never before attempted. Men who would explore the Moon – women were not selected as astronauts until 1978 – typically arrived at NASA as trained pilots, engineers, and later, scientists. Few had more than passing experience with cameras as family photographers. NASA provided professional photographers to train astronauts in composition, lighting and perspective. Astronauts learned the technical and scientific capabilities of photography, but were also encouraged to capture targets of opportunity, those awe-inspiring sights that might catch the eye of any tourist. Engineers prepared astronauts in advance with systematic plans for the photographic work while also ensuring the plans accounted for the needs of multiple audiences – NASA managers, engineers, scientists, government administrators, NASA public affairs, the media, corporations, and of course, the public.

NASA developed detailed scripts (checklists) for crews to follow throughout their missions, with photographic work spelled out and time allotted for spontaneous crew-selected images. Once images arrived back on Earth, NASA public affairs staff collaborated with the astronauts to decide which images to release publicly, and how to interpret them with accompanying text. This process, from planning, to execution, to circulation, mirrors the rigour with which NASA performed many of its public service components, making it a routine, but one that aided our understanding of one of the most stunning human achievements to date.

Apollo photography was shaped as much by the needs of Cold War politics as it was by the cultural shift of the Sixties that increased public consumption of visual media. Before, during and after the Apollo missions of the

The so-called Blue Marble *image taken by the Apollo 17 crew just five hours into their mission to the Moon, December 1972*

late Sixties and early Seventies, it was hardly possible to miss astronaut photographs, scenes of them in training or at home, on the cover of virtually any printed magazine or newspaper. But by the end of the Apollo era, those images were pushed farther and farther back in the issues, no longer gracing front covers. Traditional breaking news stories began to hold sway in editorial rooms around the country, with stories of social and economic strife, racial discord and the war in Vietnam overriding interest in space travel. Lunar missions, even by the time of the Apollo 13 emergency in April 1970, were already a minor story in the lives of most Americans. One of the most iconic images of the programme, the so-called *Blue Marble* was only published on the front page of two major US newspapers after the conclusion of Apollo 17 in December 1972: half-page sized in *The New York Times* and full-page sized in *The Boston Globe.*[2]

Photographs mediate memories of shared experiences today more than ever, as they pervade nearly every aspect of our lives. Perhaps more importantly in the case of the Apollo images, fifty years past now, the photographs create what historian Alison Landsberg describes as 'prosthetic memories'. Landsberg defines these as substitute memories developed in relation to events for which the experience was not one's own and is supported by other information about the event to reinforce existing collective memories.[3] Photographs serve to fill holes in the construction of shared identities by telling stories in and of themselves, providing

Buzz Aldrin standing on the footpad of the Lunar Module Eagle, just prior to describing the Moon as 'magnificent desolation' to Mission Control, 21 July 1969

information that goes beyond personal experience. They document human spaceflight and promote a perceived exceptionalism shaped by the accomplishments of the Apollo programme. They provide a platform for countless social and cultural byproducts never imagined by their creators. Apollo astronauts served as our proxies while carrying out their duty, capturing high-quality images that were both informative and provocative.

While many Apollo images achieved iconic status, others provide expected elements and truly exceptional moments all in a single frame. A space-suited astronaut, characteristically but casually standing next to an American flag is just one part of an image that contains more than just technological and nationalistic symbols. We can also see Gene Cernan reflected in the suit's gold-plated visor; a fragile blue orb hanging in a black void of space; all of human history (other than Command Module Pilot Ron Evans, orbiting the Moon) laid gently in one frame of 70 mm film. The sublimity, when seen as a whole, is deeply compelling.

When closely examined, images chock full of technological and scientific information on the Apollo programme document our own imprints on history. A rolling lunar landscape made a stunningly dramatic backdrop for the spidery lunar module, a parked lunar rover, and a suited astronaut setting up an experiment. Technology deflects attention from simple and profound features in the image: boot prints and tire tracks crisscrossing the powdery grey regolith. Even today, orbiting scientific satellites show us these undisturbed traces of human activity. Unlike footprints on a sandy beach washed away by the surf, these literal human impressions may always exist as artifacts of our exploration.

Ultimately, these visual documents mediate the public memory of early spaceflight: what we understand or remember about human life in outer space is largely due to the nature and character of the visuals that astronauts created. Our memories have been constructed collectively from these legacy images, shaped by priorities and procedures born of social, political and scientific goals. Questions of how and why the photographs were created, and by whom, offer viewers deeper insights into their significance as portals into this time in human history. Astronaut photography provides a particularly relevant way of examining how and why the public still uses images as a means of accessing those collective memories. Deep readings of astronaut photographs reveal them as far more than pretty pictures: their visual character embedded them in our collective memory as touchstones for the heroic era of human spaceflight.

Gene Cernan's photograph of Lunar Module Pilot and geologist Harrison Schmitt, December 1972

The Apollo 16 lunar module Orion, *Mission Commander John Young, and the crew's lunar rover on the Descartes Highlands in April 1972*

Revealing the True Colours of the Moon

Tom Kerss

We often describe the Moon in colourful terms, though little is said of its true colour. The rising or setting Moon takes on a golden shade, as sunlight reflected from its surface is filtered through the Earth's atmosphere. The so-called 'Blood Moon' takes on its eponymous deep red hue during a lunar eclipse, as the Earth's shadow is flooded with the light of every sunset and sunrise simultaneously. A 'Blue Moon', on the other hand, shows no unusual colour at all – its curious name is a traditional title from the North American farming community. But some parts of the Moon really are blue, and others appear somewhat rusty orange or brown. Indeed, you can discover this yourself with some elementary astrophotography, by exploiting the limits of digital information.

Before we see how this is done, we should acknowledge the long history of colour observation on the Moon. Subtle though they are, hints of the colours were recognised visually well before modern photographic techniques enhanced them (see opposite). To experienced moongazers, there are familiar regions where contrasting colours in close proximity pop out in the otherwise grey scene. The boundary of Mare Serenitatis (Sea of Serenity) and Mare Tranquillitatis (Sea of Tranquillity) is one such place. The former is rusty in colour, whereas the latter is a dark ocean blue. At first glance, this appears to be a meeting of two shades of grey, but careful and patient study by eye shows extremely pale apparitions of colour. More obvious perhaps is the brownish tint of Palus Somni (Marsh of Sleep) on the eastern edge of Mare Tranquillitatis. Its rugged terrain is apparently warmer in tone than the adjacent blue plains of the lunar mare, but also much brighter, and well defined by its roughness. To the eye it appears a pale brown, certainly less grey than the highlands to the east of it, which begin with the striking bright rays of Proclus crater.

Of all the colourful regions to hunt down, the Aristarchus plateau is widely considered the most conspicuous. This enormous block of highland terrain rises above the volcanic plains of Oceanus Procellarum (Ocean of Storms) on the western reaches of the Moon. Peppered with a layer of dark pyroclastic glass, a product of volcanic eruptions, the plateau nevertheless presents a strong yellow colour, sometimes described as 'mustard'. With photographic enhancement, its neighbour Aristarchus crater – unmistakably the brightest on the Moon – becomes a very strong counterpart, exhibiting a brilliant cyan. So why, in most ordinary circumstances, do we perceive this crater, as well as most of the Moon's surface, as being varying shades of grey?

Not only are the colours extremely subtle, falling below the sensitivity of the human eye, but we also do not have the luxury of storing and post-processing our visual memories to enhance them. Fortunately, modern digital cameras record colour information with extremely high precision. Often, we elect to save space by throwing much of this information away, saving our images as 8-bit JPEG files, capable of storing 16.7 million different colours. This sounds like a lot, but the RAW files most cameras are capable of making can represent a staggering 281 trillion colours. If we preserve the information, by shooting RAW image files, we can exploit this extreme precision to find, and greatly exaggerate, the many hues of the lunar surface.

For this example, 25 photos were taken in quick succession using a Digital Single Lens Reflex (DSLR) camera and small telescope, with a focal length of about one metre. This combination results in a good but forgiving image scale – that is, many details are visible, but it is not necessary for the telescope to be tracking (that is, compensating for the Earth's rotation) as there is sufficient space in the field for the Moon to drift through as a sequence of photos are taken. Each shot was then processed using PIPP (Planetary Image Pre-Processor) in order to centre the Moon and crop images. It is beneficial to throw away the black space around the Moon, because we are working with uncompressed data, and anything we can do to reduce file sizes will decrease processing time. The images were then 'stacked' using AutoStakkert! to produce a master file with a high signal-to-noise ratio, where signal is the real information consistently present in each frame, and noise is randomly distributed by the electronics of the sensor. This allows us to enhance the image without exacerbating the noise that would be more noticeable in a single frame. Finally, the image was lightly sharpened using the wavelets algorithm in RegiStax. This procedure is common practice for producing clean,

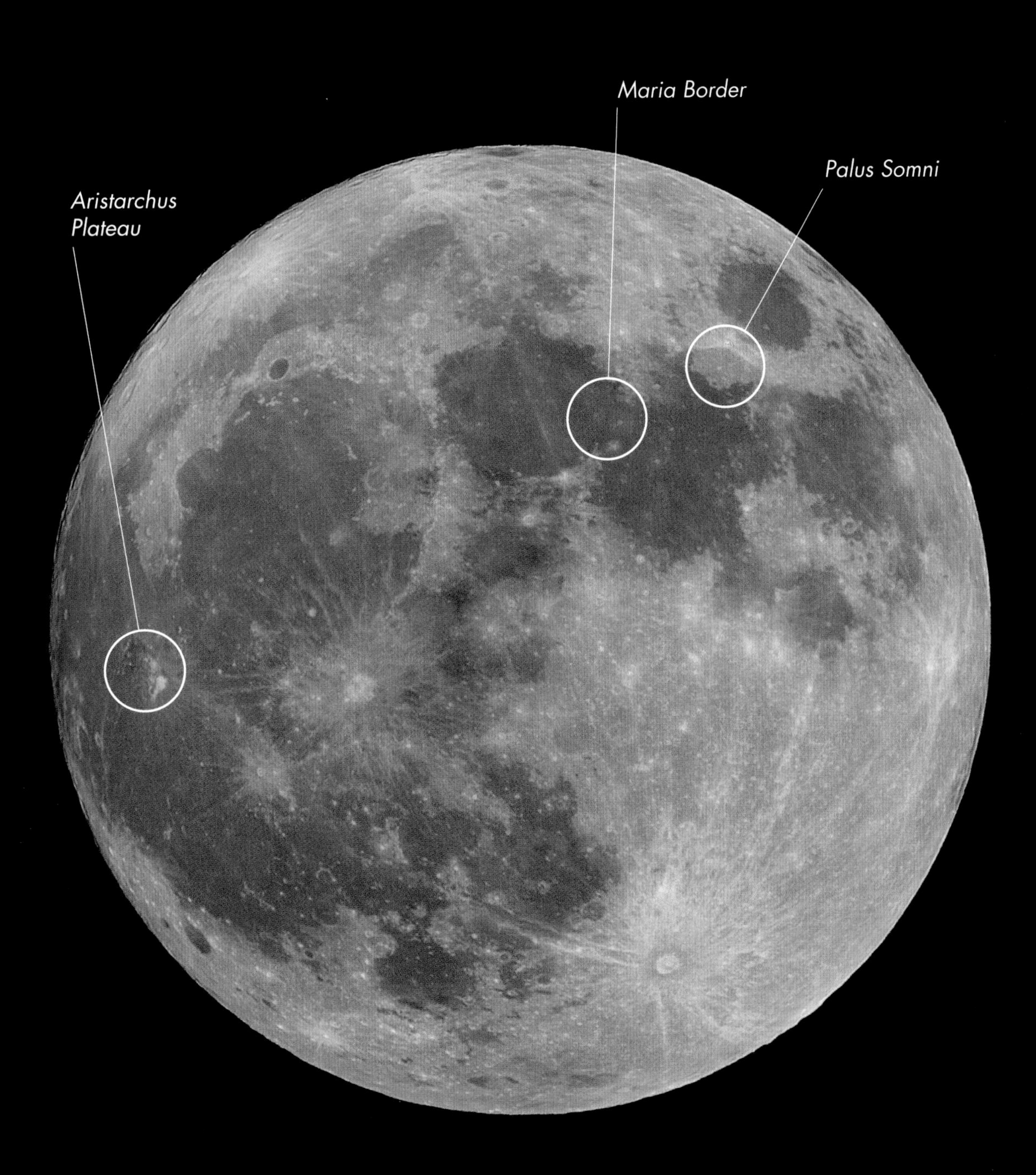

Colourful sights on the lunar surface

Stacked image before colour enhancement

Highly exaggerated colour

sharp images of astronomical objects, which are subject to distortion from the atmosphere, but it is not essential. If you have one good RAW photo of the Moon, you will be able to enhance its colour to a reasonable degree before noise spoils it. It may look something like the image top left.

Already, Mare Tranquillitatis has a pronounced, dark blue shade. This is not dissimilar to the visual view from a large telescope, and is a promising sign that there is much more to be drawn out. The colour enhancement was performed in Adobe Photoshop, using a combination of Saturation and Vibrance adjustments. Many of Photoshop's competitors also offer the same controls – the most important consideration is that the software can handle 16-bit calculations. It's advisable to make colour corrections using adjustment layers that can be hidden or switched, until the preferred aesthetic is achieved. Avoid the temptation to

Comparison between ground-based and LRO imagery

overdo it, as it is possible to introduce too much colour, or exacerbate the false colour often present in telescope optics. The predominant colours should be blue and orange or yellow. Green or violet is a sign that chromatic aberration – the misplacement of different colours as light passes through glass lenses – is leaking into the image. After some tuning, your image may look more like the image top right on the facing page.

Suddenly, our familiar satellite has a whole new personality! The cool plains and warm highlands are joined by occasional highlights of cyan in the rays of relatively recent impact craters. Realising that this colour has always been present, albeit unseen, is one thing, and producing your own beautiful photo is another. But there's a third facet to this exercise. Astronomers have been studying these colours for decades, and recently NASA's *Lunar Reconnaissance Orbiter* (LRO) – one of the most advanced satellites ever sent to the Moon – has produced a global colour map of the surface, a small section of which is seen at the bottom of the facing page in comparison with our ground-based image, confirming that the colours we see are real.

These colours actually hold great scientific value, offering clues as to the mineral composition of the lunar surface. The obvious blue tones are produced when sunlight is reflected off basalts containing relatively high concentrations (>7 per cent by weight) of titanium dioxide (TiO_2). Meanwhile, the yellow or orange hues are a combination of terrestrial-type basalts (where the titanium dioxide concentration is less than 2 per cent by weight) and a relatively high abundance of a more familiar oxide, iron oxide (FeO). High concentrations of iron oxide and titanium dioxide render the maria darker than the highlands, due to their relatively low reflectivity. The rugged older highlands with their numerous impact craters have a broadly rusty appearance, with brighter near-white streaks where relatively fresh material from more recent impacts has settled on top of the darker rock below. Colours have been used to evaluate the lunar surface and piece together its entire history.

Back on the amateur side, astrophotographers have been improving methods to present the Moon's colours from the ground. The latest techniques involve combining colour information from a DSLR or dedicated astronomy camera with ultra-crisp monochrome images of the surface details, often captured at a much higher resolution (see image below). Such images offer the illusion of more colour detail, as well as greater control in processing. The author is continuing to develop his own skills with this method, and invites you to get started in capturing the Moon's true colours. Good luck, and clear skies!

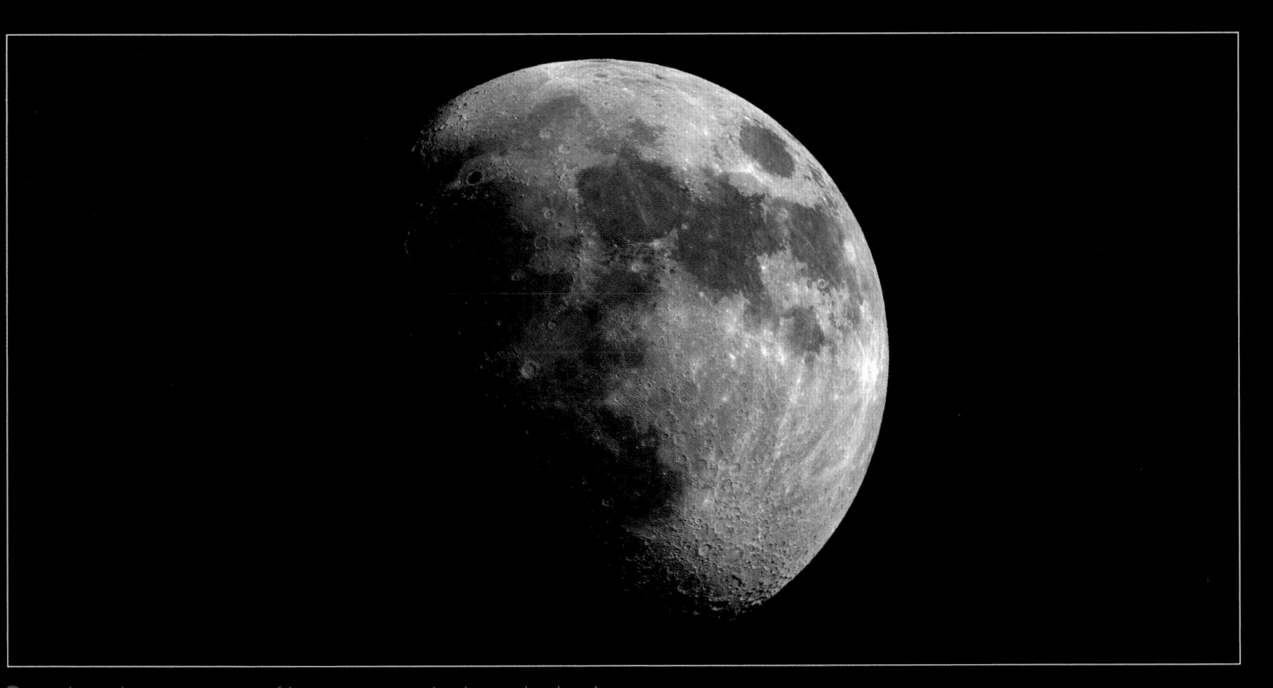

High resolution composite of lunar mosaic and enhanced colour layer

DESTINATION MOON

In 1969, Belgian graphic novel author, Hergé, sent a congratulatory cartoon drawing to Neil Armstrong of Tintin welcoming the Apollo astronaut as he stepped off the *Eagle* lunar module. Tintin had reached the Moon in 1953, in an endeavour that was both technologically plausible and told with scientific credibility. Science fiction and popular culture were way ahead of the American space programme.

For centuries, and in fact long before the Moon had been scrutinised through the lenses of a telescope, the Moon had been a destination in the human imagination. It had been a place to which things lost on Earth found their way, a safe space for satire, the home of strange creatures and otherworldly landscapes, a site to project our dreams and aspirations.

As efforts to finally reach its surface gained momentum during the Space Race in the Fifties and Sixties, the Moon permeated the domestic realm. At long last, on 21 July 1969, humanity held its collective breath as radio and TV transmitters across the globe broadcast a man's first steps on another world. As W.H. Auden wrote in his poem *Moon Landing*: 'from the moment the first flint was flaked this landing was merely a matter of time'.

Melanie Vandenbrouck

The End of the Beginning: Imagining the Lunar Voyage

Richard Dunn

People were going to the Moon long before they could get there. Imagining a voyage to our tantalisingly close yet frustratingly distant neighbour could be a way of thinking about weighty political, philosophical and religious issues. Or it could be gloriously and riotously fun. While there is much one could say about what people expected to find if they ever reached the Moon, this essay explores how exactly they thought they would get there ... and hopefully back again. Some of their tales were self-conscious flights of fancy, but others looked to the technology of their time to imagine a bit more seriously how people might one day land on the lunar surface.

Reaching for the Moon

By ancient Greek times, philosophers had begun to speculate that the Moon might be a world with inhabitants. The first surviving texts to imagine people travelling there were written around 170 CE by a Greco-Syrian satirist and rhetorician, Lucian of Samosata (*c.*125–after 180 CE). The *Alēthē diēgēmata* or *Vera historia* (True Story) is a parody of what was by then a popular genre of eventful travellers' tales; it begins by warning readers that everything that follows is a lie. It tells of a sea voyage across the Atlantic, during which the ship is caught up by a whirlwind and taken to the Moon, where the travellers find themselves in the middle of a war between the kings of the Moon and Sun. It's a brief extra-terrestrial sojourn in an otherwise Earth-bound series of over-the-top adventures.

Lucian's *Ikaromenippos* details a more deliberate lunar journey as part of a bold swipe at philosophers (arguing that their inability to agree with each other undermines their claim that philosophy can reveal unique truths). The work is a dialogue in which Menippos tells his friend how he has fashioned a pair of wings – the left taken from a vulture, the right from an eagle – and launched himself from Mount Olympus to fly to the heavens and consult with Zeus. Looking back to Earth from the Moon, he sees a tiny, fragile world whose wars, empires and philosophers' disputes seem trivial.

Lucian's two works are a good starting point because much about them would be repeated up to the Apollo era. Like Lucian, many authors described journeys to the Moon within the tallest of tales. In Ludovico Ariosto's epic poem, *Orlando Furioso* (1532), the knight Astolpho rides to the Moon in a horse-drawn chariot to find Orlando's lost wits, as the Moon is where lost things are to be found. Some authors used their lunar tales to poke fun at the foolish ideas of others, as Lucian did in his attack on philosophers. But these stories might also be bound by, and are in dialogue with, the technological possibilities of the time. To Lucian, ships were the obvious vessels in which to make long journeys, and birds the model for how one would fly. For later authors, other possibilities opened up as new technologies were developed, although there were always those who gleefully ignored the constraints of realism.

Moon travel really took off in the seventeenth century. In 1608, the astronomer Johannes Kepler (1571–1630) wrote a speculative and surprisingly fantastical work, *Somnium* (The Dream), which circulated in manuscript before being published in its final form in 1634. It describes a dream in which daemons carry an Icelandic boy along a pathway to the Moon that is only passable during a solar eclipse. Here though, the means of travel is secondary in a work more concerned with the nature of the Moon, including its elaborate flora and fauna, the extremes of heat and cold, and what the Earth might look like from its surface. Copious footnotes also detail the astronomical and other data on which the more imaginative dream-narrative is based.

All things lunar changed with the invention of the telescope and the publication in 1610 of Galileo Galilei's *Sidereus Nuncius*, which outlined shocking claims including the discovery of mountains on the Moon. If this was true, the Moon was an Earth-like world to which one could truly contemplate journeying. Such ideas inspired a host of lunar travels drawing on ideas old and new. English playwright Ben Jonson penned a satirical masque, *News from the New World Discover'd in the Moone* (1620), in which the telescope allows earthlings to communicate with

Travelling to the Moon in a chariot, illustration by Gustave Doré, from Ludovico Ariosto, Roland furieux *(a French edition of* Orlando Furioso*) (Paris: Hachette et Cie, 1879), p. 443*

their lunar counterparts. It also mentions three ways one might get there: with wings, as Menippos did; in sleep or a dream, like Endymion, who was visited each night by Selene, goddess of the Moon; or by leaping into a volcano and being carried Moon-wards by the force of an eruption, as the philosopher Empedocles was said to have done. In *The Man in the Moone* (1638), English bishop Francis Godwin (1562–1633) describes the travels of Spanish merchant Domingo Gonsales. Hoping to return home from the Atlantic island of St Helena, Gonsales builds a flying machine drawn by 'gansas', wild swans or geese. But instead of taking him home, they fly him to the Moon (in under 12 days), where he discovers a utopian society of devout Lunars. Godwin, who notes Galileo's telescopic discoveries, uses this speculative fancy to explore the theological implications of extra-terrestrial life, including the possible salvation of alien souls.

Godwin's work had some influence at the time and still inspires artists today (as Melanie Vandenbrouck discusses in 'Whose Moon? The Artists and the Moon' later in this volume). Having published *The Discovery of a World in the Moone* (1638), another text on the religious implications of life on other worlds, Puritan chaplain John Wilkins (1614–72) took Godwin's ideas on board in a second edition of 1640. Acknowledging Gonsales's tale, Wilkins adds his thoughts on possible means of lunar travel. These include using wings strapped to one's body, being transported by birds like Gonsales or building some form of mechanical winged chariot. If this last idea were perfected, he notes, it would revolutionise travel on Earth as well.

A more light-hearted reference to Godwin appears in the parodic *Histoire Comique, par Monsieur de Cyrano Bergerac Contenant les Estats & Empires de la Lune* (1657) (Comical History by Cyrano de Bergerac Containing the States and Empires of the Moon). Written as if it were the personal experiences of the writer and soldier Savinien de Cyrano de Bergerac (1619–55), the book tells of Cyrano's quest to discover whether the Moon is inhabited. Getting there proves tricky, however. First, he straps vials of dew to his body so that they will lift him up as the dew is drawn up by the Sun's heat. When this fails, taking him only as far as Canada, he builds a flying machine that he launches from a cliff. It crashes but then some soldiers attach rockets to it to make it fly at a religious celebration. While Cyrano is trying to remove the rockets, they ignite and the machine takes off with him in it, transporting him to the Moon, where he meets its inhabitants and, among others, the prophet Elijah and Godwin's Domingo Gonsales. An alternative proposal appears in *Iter Lunare: Or, A Voyage to the Moon* (1703), a critique by English ship chandler David Russen of previous works including those of Godwin and Cyrano. Interrogating the scientific credibility of these tales, Russen imagines instead a gigantic spring-powered catapult used to launch the prospective lunar traveller into space.

Other writers followed Cyrano's fantastical tradition. The most famously outlandish stories were those of Baron Munchausen, a fictional nobleman and teller of tall tales based on the real (but highly unimpressed) Karl Friedrich Hieronymus, Freiherr von Münchhausen. German writer Rudolf Erich Raspe (1736–94) began publishing Munchausen's impossible adventures from the early 1780s. Two of these take his hero to the Moon. In the first, the baron climbs a beanstalk to retrieve a hatchet he has accidentally thrown there. When the giant plant withers as the Sun comes up, Munchausen makes a rope from straw and lets himself down by repeatedly cutting it at the top, lowering it and re-tying it at the bottom to get back to Earth. The second journey recalls Lucian, as a whirlwind drives the baron's ship to the Moon, whose inhabitants ride three-headed birds and use radishes as weapons.

Marvellous machines and strange substances

What should have become clear by now is how easily the Moon journey became a vehicle for comedy and satire. Some of these satires depicted ridiculous schemes that were simply enlarged versions of mundane objects such as ladders or bridges. While these satirical notions continued to infuse tales of lunar voyaging, new inventions and discoveries seeped into writers' imaginations alongside them. Around the time Raspe was publishing Munchausen's tales, Europe was succumbing to a new craze following the Montgolfier brothers' public demonstration of a crewed balloon flight in June 1783. It was seized upon in *Le Char Volant, ou, Voyage dans la lune* (The Flying Chariot, or a Trip to the Moon, 1783) by Cornélie Wouters, Baroness de Vasse, whose Moon is a feminine utopia to contrast the patriarchal misery of Earth.

More lunar balloonists followed in Jacques-Antoine Dulaure's *Le retour de mon pauvre oncle, ou Relation de son voyage dans la Lune* (My Poor Uncle's Return, or an Account of his Trip to the Moon, 1784) and Daniel Moore's *An Account of Count D'Artois and His Friend's Passage to the Moon* (1785). The pseudonymous Aratus's *A Voyage to the Moon, Strongly Recommended to All Lovers of Real*

Harnessed 'gansas' carry Domingo Gonsales aloft, from Francis Godwin, The Man in the Moone *(2nd edition, London: Printed for Joshua Kirton, 1657), frontispiece*

Cyrano de Bergerac attempts to fly to the Moon, powered by vials of dew, from Cyrano de Bergerac, ΣΕΛΗΝΑΡΚΙΑ [Selēnarkia], or, The Government of the World in the Moon: A Comical History *(London: J. Cottrel, 1659), frontispiece*

Baron Munchausen returns to Earth (far right), from *The Surprising Travels and Adventures of Baron Munchausen* (London, c. 1840)

Freedom (1793) is a revolutionary call to arms. Half a century later, the American writer of the mysterious and macabre, Edgar Allan Poe (1809–49), made his mark with 'Hans Phaall – A Tale', published in the *Southern Literary Messenger* in June 1835 and in a fuller version five years later as 'The Unparalleled Adventure of One Hans Pfaall'. Poe's hero, a mender of bellows, builds a balloon filled with a gas much lighter than hydrogen in order to escape his creditors. It takes him to the Moon in 19 days.

Other technological advances soon found their place. Appropriately for the age, a steam-driven spaceship features in *The Marvellous and Incredible Adventures of Charles Thunderbolt on the Moon* (1851), a children's story by Charles Rumball (writing as Charles Delorme). Jules Verne (1828–1905), one of the most influential nineteenth-century writers of extraordinary voyages, looked to advances in artillery science, although many considered this unworkable due to the potentially fatal acceleration a gun-like explosion was expected to cause. Nevertheless, in *De la terre à la lune* (From the Earth to the Moon, 1865), Verne meticulously sets out the scientific and technical details of just such a Moon shot, describing a mammoth endeavour requiring committee meetings, sandwiches and a great deal of money. The proposed spaceship, built at the instigation of the Baltimore Gun Club, is fired from a huge cannon sunk within a mountain. The grandiose plan excites first America and then the whole world, with only Britain – the Royal Observatory included – proving unsupportive due to national jealousy. Much has since been made of the proximity of the suggested launch site in Florida to that used for Apollo a century later, but what is more notable is the imperialist, conquering ambition of Verne's American protagonists. In fact, the book only gets as far as the building of the aluminium ship and its firing into space, the destiny of its occupants left uncertain. Readers finally discovered their fate five years later in *Autour de la lune* (Around the Moon, 1870). Having survived the explosive launch, the travellers face an eventful journey, during which they have to dispose of the corpse of one of their dogs, experience weightlessness, talk science and narrowly avoid oxygen poisoning. They reach the Moon but are only able to orbit, having been sent off course by the close pass of an asteroid. Eventually managing to return to Earth, they splash down in the Pacific to be rescued by ship.

Advances in physics and chemistry, including the isolation of new elements, gave inspiration too. Several

'The projectile passing the Moon' from Jules Verne's The Moon Voyage (London: Ward Lock, 1877)

stories invoked imaginary materials with anti-gravitational properties. Writing as Joseph Atterley, George Tucker put forward 'lunarium', a metal that repels Earth but is attracted to the Moon. *A Voyage to the Moon* (1827) describes an airtight vessel made of the amazing substance, the basis for a satirical travelogue. In *Orrin Lindsay's Plan of Aerial Navigation* (1847), John L. Riddell conjures up an amalgam of mercury and steel that counteracts the Earth's pull when placed in a magnetic field. Lindsay builds a ship covered in the amalgam, using shutters to vary the forces as needed. In tune with the scientific thinking of the time, however, his Moon is devoid of life. Chrysostom Trueman's *A History of a Voyage to the Moon* (1864), by contrast, imagines an ideal lunar society of reincarnated humans discovered by a ship powered with 'repellante', an anti-gravity ore mined in the Colorado mountains.

Anti-gravity was also the means deployed in a tale to rival Verne's in terms of popularity and influence. *The First Men in the Moon* by H.G. Wells (1866–1946) first appeared as a serial in *The Strand Magazine* between December 1900 and August 1901, then as a book a few months later. Wells's magical material is a metallic substance called 'cavorite' after its inventor, a chemist named Cavor, who uses it to make a spherical spaceship with sliding windows or blinds that allow it to be steered.

On the journey into space, Cavor and the narrator, an entrepreneur named Bedford, experience the weightlessness we now inevitably associate with space travel, the disorientation of looking down onto the Earth and the startling brilliance of the Moon. Having landed, they encounter an advanced race of insectoid Selenites living beneath the surface, who take the two men captive. Bedford manages to escape and return to Earth, but Cavor is forced to remain. Later radio transmissions reveal that the scientist has survived and learned much from his captors, but communication ceases once the Selenites come to understand humanity's propensity for war. While certainly a tale of off-world adventure, *The First Men in the Moon* is also a critique of Victorian imperialism and unthinking attitudes towards technological progress. Through Cavor, Wells warns of the consequences of scientific curiosity untempered by social conscience; through Bedford, of the dangers of unfettered commercialism.

The popularity of lunar voyaging by the turn of the twentieth century saw imaginative journeys in other spectacular forms. Frederick Thompson staged 'A Trip to the Moon' at the 1901 Pan-American Exposition in Buffalo. Visitors took a spaceship, *Luna*, which 'flew' past artificial clouds before models of Buffalo and the Earth appeared below. Landing on the Moon's cratered surface, passengers took a stroll then returned home. The ride later became the centrepiece of the Luna Park on Coney Island, New York. The latest entertainment form, moving pictures, also latched onto space travel. French cinematic pioneer George Méliès (1861–1938) famously mashed Verne and Wells together in the fanciful and comical *Le Voyage dans la Lune* (A Trip to the Moon, 1902), in which a group of scientists travel to the Moon in a cannon-propelled capsule, explore, fight and return to Earth with a captured Selenite.

Chiefly remembered for the inventiveness of its visual effects, *Voyage dans la Lune* nevertheless hints at anti-imperialist sentiments in its harsh image of the Selenite prisoner paraded before baying crowds. In a rather

'Provisioning the wonderful sphere' and 'Sunrise on the Moon', from a screen adaptation of The First Men in the Moon *(Gaumont Film Hire Service, London, 1919)*

Le Voyage dans la Lune *(Georges Méliès, France, 1902)*

Firing the cannon in Le Voyage dans la Lune *(Georges Méliès, France, 1902)*

different but no less comic film, *The '?' Motorist* (Walter R. Booth, UK, 1906) goes on the run from the police, taking his automobile – another recent invention – for a spin around the Moon and Saturn's rings. Wells continued inspiring filmmakers, with the first adaptation of *The First Men in the Moon* released in 1919, although it is now sadly lost.

The race to space

Verne was apparently critical of Wells's scientifically implausible cavorite, although for Wells the science was not the heart of the story, rather the means to different narrative ends. Other authors, however, were keen to seize upon new technological possibilities from the real world and might themselves be at the cutting edge of the latest research. Russian teacher and scientist Konstantin Tsiolkovsky (1857–1935) began his career as one of a small number of people thinking about rocket propulsion. He wrote papers on multi-stage rockets and other ideas, including what would turn out to be the crucial suggestion of using liquid propellants, but the publication in which some of his proposals first appeared was banned by the secret police. He turned instead to fiction, writing *Vne Zemli* (Outside the Earth), in 1916, with a fuller version in 1920, in which an international team of scientists builds a space rocket. It takes them to the Moon, where they explore the far side in a buggy and discover plant life that can move around.

Near the end of his life, Tsiolkovsky was a consultant on the first Soviet feature film about a Moon voyage, *Kosmicheskii reis* (Cosmic Voyage, Vasily Zhuravlyov, Soviet Union, 1936), which naturally featured a rocket-ship. In combining the roles of scientist, writer and consultant, Tsiolkovsky reflected a productive relationship that underpinned a number of fictional lunar travels in the decades leading up to the Apollo 11 Moon landing. Such narratives were increasingly inflected with the international rivalry of the Space Race. Two films made either side of the Second World War illustrate this well. The first, *Frau im Mond* (Woman in the Moon, Fritz Lang, Germany, 1929) can reasonably claim to be the earliest cinematic attempt to depict spaceflight realistically, thanks to technical advice from Hermann Oberth (1894–1989) and Willy Ley (1906–69). Born in what is now Romania, Oberth had become obsessed with space flight from reading Jules Verne as a child and in 1923 published *Die Rakete zu den Planetenräumen* (The Rocket into Planetary Space), which acknowledged his debt to Tsiolkovsky's pioneering work. Though complex and controversial, the book inspired both Ley, who would become an influential science writer and a keen propagandist for rocketry, and Wernher von Braun (1912–77), the leading figure in Nazi Germany's missile programme during the Second World War and then in the American space programme after 1945.

The determined advocacy of Ley and others sparked something of a rocket fad in Berlin and *Frau im Mond* marked its highpoint. As technical adviser, Ley helped shape a film that leans heavily towards the practical details of getting a rocket to the Moon, although there is considerable melodrama – gold and greed, peril and romance – once the rocketeers arrive there. But it is the technical details that stand out in the long and painstaking preparations for lift-off, and certain elements now appear strangely prescient: the countdown to zero (Lang's idea to add a bit of drama); the separation of the booster rocket (designed by Oberth); the effects of gravity at lift-off and weightlessness in space. Sadly, Oberth failed in his plans to build a working rocket for the film's premiere. Yet the film's apparent authenticity was sufficient for the Gestapo to confiscate the production models of the rocket in 1937 and withdraw the film from overseas circulation at a time when Germany's covert rocket programme was well underway.

Though made in different circumstances, *Destination Moon* (Irving Pichel, USA, 1950) shares many characteristics with Lang's film, in particular the emphasis on the preparations and flight, which take up two-thirds of its 90 minutes. Like *Frau im Mond*, the shorter section on the Moon is somewhat melodramatic, with a crewman narrowly avoiding being marooned. The film's adviser was science fiction author Robert A. Heinlein (1907–88), on whose *Rocket Ship Galileo* (1947) the screenplay was loosely based. Having previously worked in aeronautics for the US Navy, Heinlein brought his experience to a film that flaunts its authenticity. Publicity boasted that *Destination Moon* showed how space travel would be in just a few years, something echoed in its closing words on screen, 'This is THE END of the Beginning'. The supposed realism of the moonscapes by renowned space artist Chesley Bonestell was bound up in the same hype. But *Destination Moon* also bears the hallmarks of the Space Race, which had been gaining pace since the end of the war. The film articulated the concerns of a vigorous campaign to encourage the US administration to expand its rocket programme in the face of foreign (i.e. Soviet) competition. 'He who controls the Moon controls Earth', the rhetoric made clear.[1] In a notable contrast to the way in which space programmes were being

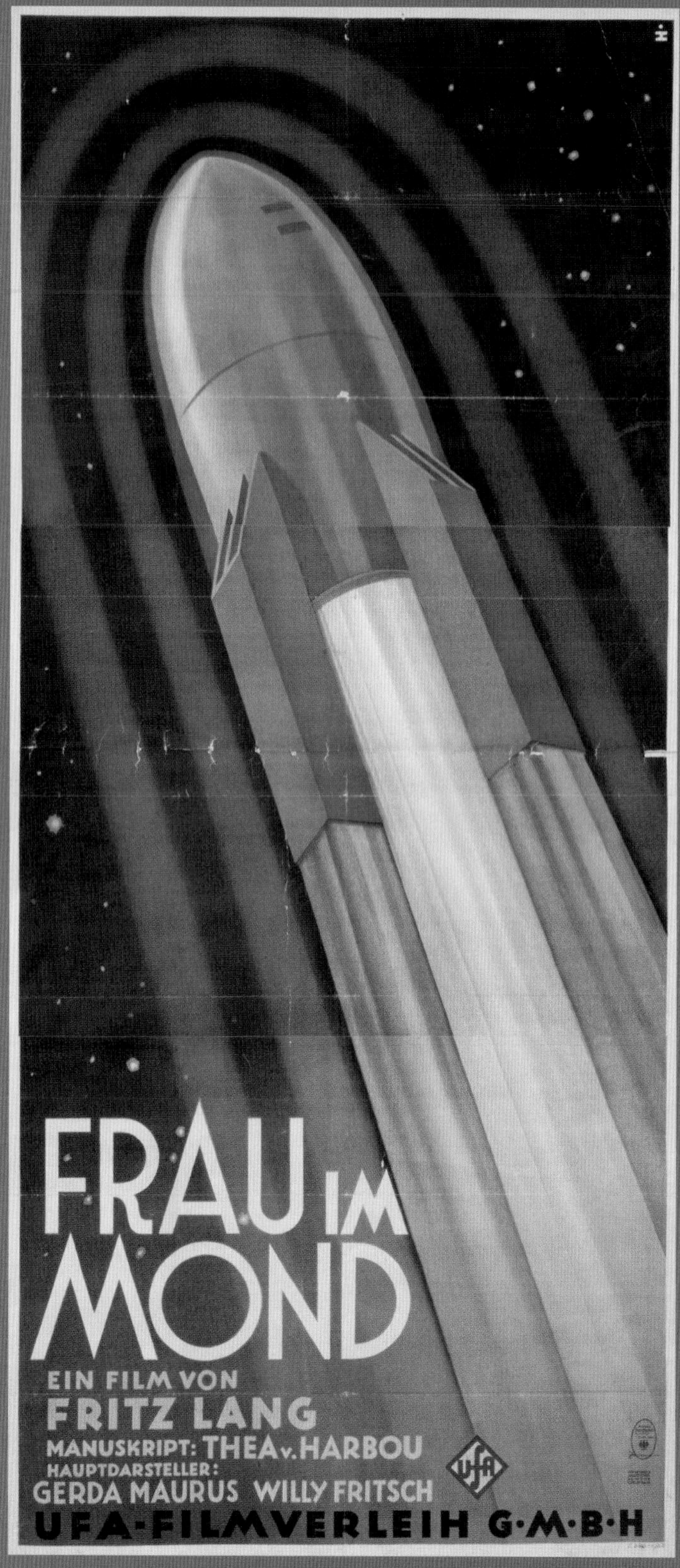

Poster by Herdmann for Frau im Mond *(Fritz Lang, Germany, 1929)*

funded at the time, however, it is American industrialists (encouraged by cartoon character Woody Woodpecker), rather than the state, who ensure that the Moon rocket is built. Perhaps this is the film's most uncanny precedent for the twenty-first century.

Destination Moon marked the start of over a decade in which travelling to the Moon with rockets for exploration rather than destruction seemed inevitable. Even Tintin went there (in an atomic rocket). Yet an undercurrent of international rivalry persisted, even more so after the Soviet Union's game-changing launch of *Sputnik 1* in 1957. A second film version of Wells's *First Men in the Moon* (Nathan Juran, UK, 1964) played with this rivalry, framing the story as the discovery by a pioneering American mission that British astronauts had already been there decades before. But there was also an important creative dialogue at work. Film and TV productions increasingly sought technical advice from the scientists and engineers leading the race to space. Now heading up the US rocket programme, von Braun became a prominent face on screen, fronting a three-part series on space travel for Disney and advising on *Conquest of Space* (Byron Haskin, USA, 1955), another film that boasted its scientific credentials. By the Sixties, space travel would be commonplace, with established iconographies: multi-stage rockets belching smoke at lift-off; cratered lunar landscapes; intrepid astronauts venturing into the unknown. Yet, as Simon Guerrier discusses in his essay, it would all change when the *Eagle* landed.

Dr Charles Cargraves and American industrialists discuss the proposed rocket project; production still from Destination Moon *(Irving Pichel, USA, 1950)*

 Project Moon: Satirising our Satellite

Katy Barrett

In *Orlando Furioso* (1516, first published 1532), the Italian poet Ludovico Ariosto sent his character Astolfo to the Moon by chariot. There, Astolfo found men and women searching for things lost on Earth:

> *One, while he loves; one, seeking fame to gain;*
> *One, wealth pursuing through the stormy main;*
> *One, trusting to the hopes which great men raise,*
> *One, whom some scheme of magic guile betrays.*
> *Some, from their wits for fond pursuits depart,*
> *For jewels, paintings, and the works of art.*
> *Of poets' wits, in airy visions lost,*
> *Great store he read; of those who to their cost*
> *The wandering maze of sophistry pursu'd;*
> *And those who vain presaging planets view'd.*[1]

Ariosto's was one of the first works to imagine a voyage to the Moon and influenced a host of later European dreamers and satirists. While his Moon was a perfect orb in the sky, home to the 'man in the moon' or the goddess Luna, with the invention of the telescope in the early seventeenth century it became a solid body with a visible, and potentially habitable, surface. Greater scientific understanding of the Moon went hand-in-hand with its continued popularity as a site for fiction and satire.

Orlando Furioso was first translated into English in the late eighteenth century by John Hoole, whose rendering nicely captured the range of human desires and fears with which the Moon continued to be associated, from unrequited love and the search for fame, to speculation on others' inventions and, of course, madness. This period was also the height of English visual satire. From William Hogarth in the early decades of the eighteenth century, to William Blake at the start of the nineteenth, artists commented on and questioned their society in printed images that had a wide dissemination. The Moon offered a number of ways to question life on Earth, but its role could be condensed to the idea of 'a project': a naïve, over-ambitious or even malicious scheme that caused public investors to suffer. Projects could be political, financial, religious or scientific, but all had an ethereal, insubstantial quality: when investigated, they disappeared into thin air, or floated up to the Moon.

The first great failed project of the century was the 'South Sea Bubble' of 1720, in which investors lost fortunes in the crash of stocks in the South Sea Company, and the complicity of political and financial figures in the disaster became worryingly clear.[2] Hogarth was still wrestling with these revelations in 1724 when the flood of pamphlets and publicity around a lunar eclipse captured the public imagination. He chose to frame his ongoing concern at public 'projecting' within a telescope tube aimed at the Moon, and a print entitled *Some of the Principal Inhabitants of the Moon as they Were Perfectly Discovered by a Telescope brought to ye Greatest Perfection since ye last Eclipse Exactly Engraved from the Objects, whereby ye Curious may Guess at their Religion, Manners &c.*

Hogarth shows a cloudy platform, evoking the castles in the air that satirists presented investors as building during the 'bubble crisis'. On it sit a king, judge and bishop whose bodies are made up of various objects. The king's face is a coin, while his chain of office and his ceremonial orb are made of bubbles, marks of office that will burst when touched. Both the orb and his sceptre are crowned by crescent moons, ultimate signs of his changeability. Thus, the waxing and waning Moon is made representative of the instability of such financial and political projects as the South Sea Company, while the telescope's lens implies altered vision and the difference between appearance and reality.[3]

Hogarth's print was reproduced by Samuel Ireland in 1788. With the invention of the hot air balloon by the Montgolfier brothers, Joseph-Michel and Jacques-Étienne, earlier in the decade, the idea of flying to the Moon was by this time no longer beyond the realm of physical possibility. Yet, the idea of reaching for it remained connected with projects. 'Balloonomania' seized the British public's imagination to the same extent as the South Sea Bubble had, and led to a flurry of printed poems, satires and images, as well as a new fashion for hair *au demi-ballon* and even balloon-back chairs. In early 1784, the Montgolfier brothers' 'Grand Aerostatic Globe' went on

Some of the Principal Inhabitants of the Moon …
Engraved by Samuel Ireland in 1788, after William Hogarth, 1724–5

Chevalier Humguffier [Momguffier] and the Marquis de Gull making an excursion to the Moon in their new aerial Vehicle

display at the Lyceum in London, and was described with awe by the *Morning Post*:

> *This brilliant and most magnificent spectacle is doubly overlaid with gold! Upon it beam [...] constellations of stars, and all the planets of our solar system! [...] the whole exhibits the appearance of a huge world floating in the incomprehensible infinity of eternal space!*[4]

A satire published by James Basire later that month shows just such a celestial balloon taking the 'Chevalier Humguffier and the Marquis de Gull' on a voyage to the Moon.

In the example in the Science Museum collection, a manuscript edit has changed Humguffier to 'Momguffier'. It is undoubtedly intended to evoke the Montgolfiers, but is more generally a satire on ballooning as a fraud. With 'hum' as an eighteenth-century satirical term for a scam, and 'gull' referring to a cheat, ballooning becomes just another malicious project to distract the public and encourage deluded reaching for the Moon.[5]

The 'projecting' mania of the 1720s gripped Britain again in the early nineteenth century. The South Sea Crisis had led to the passing of the 'Bubble Act' in 1720, intended to prevent unincorporated companies from behaving like incorporated ones and selling shares. Although unincorporated activity did continue during the eighteenth century, the act had not been invoked since 1723 when in 1807 its terms were used to question the legality of companies promoted by the engineer Ralph Dodd. As well as advertising enterprises for paper manufacture and distilling, Dodd had developed projects for a dry tunnel under the Thames from Gravesend to Tilbury (anticipating Brunel), and for a Strand Bridge over the river (which opened as Waterloo Bridge in June 1817). Subscribers to his tunnel scheme lost their investments.[6] Dodd therefore featured as one of the characters in *The School of Projects* published in *The Satirist* on 1 October 1809.

Against a dark sky filled with smoke, five men are intent on new inventions. At the left, Dodd and an associate focus on an intricate model that shows a bridge connecting the Earth to the Moon, as they also gouge a tunnel into

The School of Projects *by Samuel de Wilde, published in Samuel Tipper's* The Satirist, or Monthly Meteor, *1 October 1809*

the Earth. The accompanying account of visiting the school is almost half taken up by Dodd's Moon project and its support by his fellow projectors, outlining how

> *Another projector had constructed a carriage to cross to the Moon which would go without horses […] the president had drawn up a plan for the insurance […] and the whole was to form one grand scheme, which, in allusion to the proposed market in the Moon, was to be entitled the 'Lunatic Company'.*[7]

The Moon is the site of projecting inventions, which will inevitably lead to madness, both of the inventor and the investing public.

William Blake is a less straightforwardly satirical figure. Originally produced in 1793, his *For Children: The Gates of Paradise*, a tale as much optimistic as cautionary about human life, was reissued in about 1820 as *For the Sexes*. Across 18 diminutive engravings (each smaller than a playing card), Blake follows the psychological course of human life from birth to death, through a series of emblems and short inscriptions. In Plate 9, at the height of man's adult emotions, a small figure reaches for the Moon, crying 'I want! I want!'.

It is an image both innocent and ambitious. Blake seems to be supporting the necessity of poetic imagination in striking a new path, at the same time as satirising the downfall of such over-reaching ambition: the original had appeared at the height of his disillusionment with the French Revolution. Indeed, Blake comments on the inventor's enthusiasm made practical. This tiny figure seeks to reach the Moon by climbing a ladder, while his companions cling to each other in concern, but it is a rocky physical Moon that he aims for, not merely the idea of one, as the ladder visibly rests on the inside edge of the crescent.[8] This is the simplest of projects, and Blake's inventor's attempt seems to endanger only his own future. Yet, Blake has put his enthusiasm at the centre of man's life cycle: our relationship with the Moon as the ultimate source of ambition, and of failure.

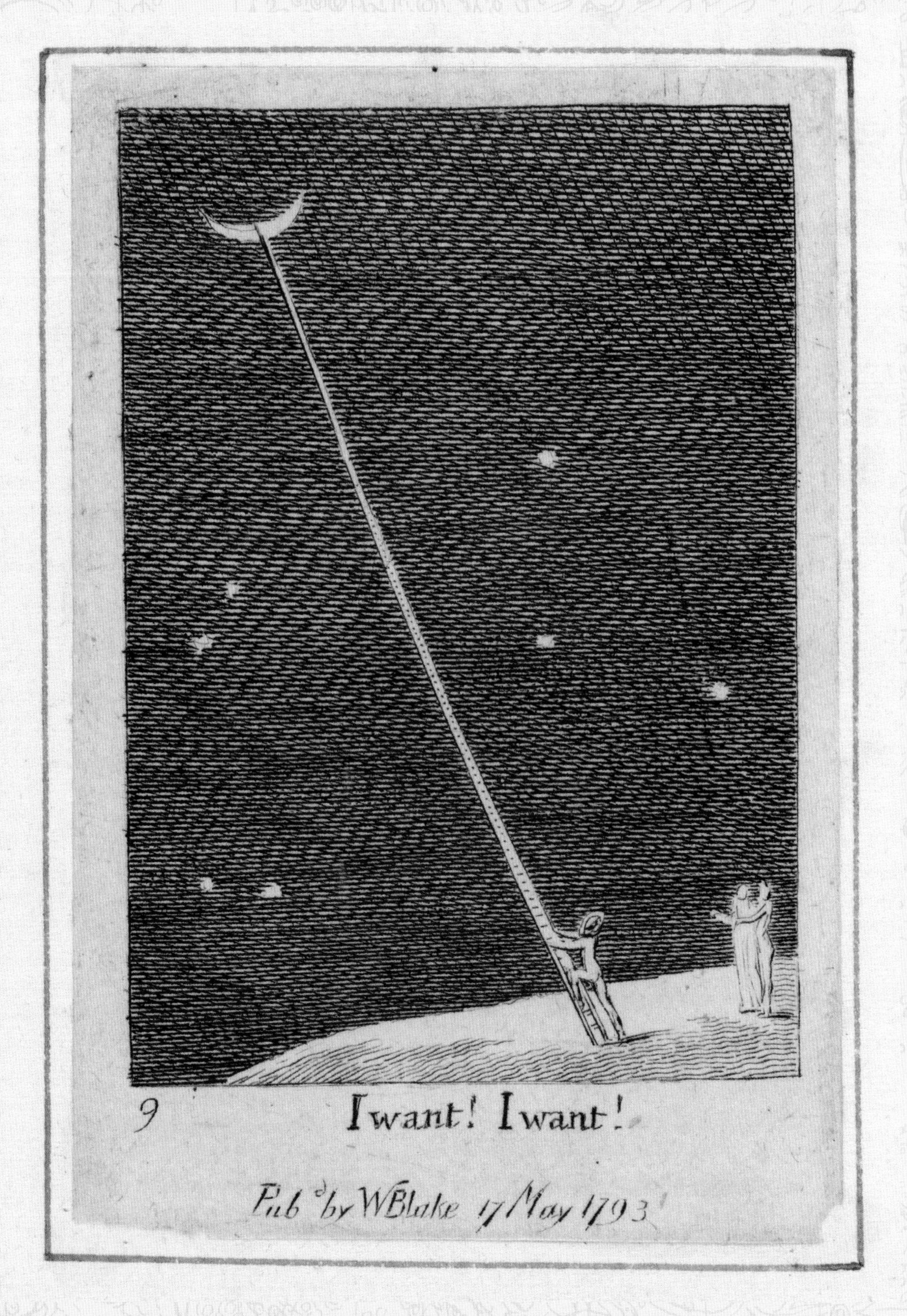

I want! I want! *by William Blake, published in* For the Sexes: The Gates of Paradise *(c.1820)*

Life on the Moon, Newspapers on Earth

James Secord

Other than the Apollo landings, the most celebrated news ever reported about the Moon appeared during the summer of 1835 in the *Sun*, a cheap newspaper for the working classes in New York. The *Sun* announced that a Scottish newspaper, just arrived with the transatlantic mail, carried extracts from a scientific journal describing the discoveries of the British astronomer Sir John Herschel during a secret mission at the Cape of Good Hope: a new kind of telescope had revealed forests and basalt columns on the Moon. The next instalment described the 'specimens of conscious existence' observed by the astronomers, including herds of diminutive bison, blue goats with a single horn, and water birds akin to pelicans and cranes.[1] The third report heralded the discovery of shellfish, birds, reindeer, elk, moose, horned bears, and (most intriguingly) biped beavers. Although similar to terrestrial beavers, these lacked tails and walked on two feet, carrying their young in their arms.

The fourth article announced the most astonishing finding of all: 'flocks of large winged creatures' akin to humans. From their graceful bearing and rapid gesticulations, the astronomers 'inferred that they were rational beings' and named them 'Vespertilio-homo', or 'bat-man'. A distinctive race, lighter in colour, was also observed, and was described (in rhetoric mirroring contemporary racial discourse) as more advanced. Other large mammals and a near-angelic bat-race were discovered later, along with a temple. These were announced in the final pair of extracts, which appeared as August drew to a close

At this point everyone who tells this classic story enthuses about an explosion of public interest. As the writer Edgar Allan Poe noted, it was 'the greatest *hit* in the way of *sensation* – of merely popular sensation – ever made by any similar fiction either in America or in Europe'.[2] Two Yale science professors are said to have travelled to the *Sun* offices to check out the story. A captain reportedly set sail for the Cape to witness the wonders of lunar life. A delegation of Baptist clergymen is supposed to have asked if there was any method for conveying the gospel to 'brethren in the newly-explored regions'.[3]

Alas, there were no Yale professors, no sea captains and probably no Baptist ministers. Although the lunar discoveries were certainly widely circulated, the *Sun* was eager to make as much out of the sensation as it could, and its selective reporting and misleading quotations continue to shape our understanding. Before we laugh at the gullibility of nineteenth-century Americans, it makes sense to examine why we have been so willing to believe such tall tales ourselves.

We can begin by recognising that 'the Moon Hoax' wasn't a hoax. The author, Richard Adams Locke, far from thinking readers would be misled, aimed to ridicule arguments for life on the Moon. His target was the Scottish astronomer Thomas Dick, a celebrated 'Christian philosopher' who atoned for the sin of seducing his maidservant by advocating divine design and the existence of extraterrestrial life. Locke launched his campaign as an anti-theological satire, attacking Dick's '*imaginative school of philosophy*' as a manifestation of religious superstition.[4]

From the moment the *Sun* hit the streets, however, no one got the point. Instead of debating the mixing of theology with science, readers were discussing whether the lunar reports were true. In America, with the populist Andrew Jackson as president, elites were under attack and scientific satire was difficult to interpret. A reader needed confidence to challenge technical-sounding descriptions, and life on other planets had a high degree of plausibility. The boundaries of the realms of writing on which the Moon story touched, from factual reporting to fictional romance, were shifting. The foundations of knowledge were uncertain, more so than at any other time since the seventeenth century.

The *Sun*'s unprecedented success provoked reactions from competing newspapers in New York. Three days after the initial report, several morning papers carried the story as far as the discovery of large animals. One expressed cautious acceptance, especially as Herschel was known to actually be at the Cape.[5] Another proclaimed that 'Sir John has added a stock of knowledge to the present age, that will immortalize his name'.[6] A third appeared late enough in the day to include that morning's instalment about

Scoperte fatte nella luna dal Sig.r Herschell *(Discoveries made in the Moon by Sr. Herschel)*, c.1835

Snuff box depicting Sir John Herschel, Germany, c.1835

bat-men, but made no comment.[7] Almost immediately, however, the tide turned against the *Sun*, one editorial joking that the articles should be bound up with *Gulliver's Travels*.[8]

Once bat-men were sighted, no New York papers gave the story credence. Outside the metropolis, a few editors were more charitable; an Albany daily welcoming the findings 'with unspeakable emotions of pleasure and astonishment'.[9] Such comments, as with Orson Welles's Mars invasion broadcast a century later, were far less widespread than one would guess. Instead, this was a charivari of popular journalism, with all the trappings of

Wallpaper fragment: 'A Peep at the Moon'

commodity production from wallpaper and satirical prints to snuffboxes and fans.

The Moon story also raised questions about expertise, authority and trust. Relying on impeccable-sounding sources, it offered a telling commentary on scientific objectivity, showing that nothing could be taken for granted. Readers needed to be on their guard; the elite papers that had supposedly swallowed the story could be revealed as corrupt, biased and incapable of training the public mind. The cheap penny newspapers saw the Moon story as a skilful joke, perhaps carried too far, but putting issues of evidence and objective reporting on the agenda. In a true democracy, the reality of facts could only be decided by a vigilant, educated citizenry.

If individuals in the egalitarian, demagogic world of Jacksonian America were supposed to judge for themselves, the situation was very different when the story reached Europe three weeks later. Republished and read as a completed narrative, rather than as serial revelations, the reports lost the ambiguity of the initial reception, so that readers could only wonder at the credulity of their transatlantic brethren. Read as an entertaining instance of human gullibility, the reports reassumed much of their intended function as an attack on religion. The London-

based radical *Satirist* claimed that the 'Ass-tronomer Herschel' had been told by the 'Autocrat of all Lunatics' of a huge structure on the Moon, rotting and falling apart, called the 'Church Establishment'.[10]

One of the most remarkable versions of the story appeared in Welsh in the 1850s. Based on a pamphlet version published in London, this was one of the unusual short works issued in a campaign for the Welsh language by John Jones of Llanwrst. In Naples, three different firms produced elaborate lithographs of the discoveries, part of a long tradition of illustrating lunar fantasies. Targeted at tourists and collectors, these images satirised the story as a parable of racial conquest in the Americas.

The events of 1835 are regularly presented as a delusion that 'swept' across the world. The 'mass public' is seen to act instantaneously and in unison. These assumptions have created the real Moon hoax: a hoax about reception, authority and readership that has been far more insidious than claims about biped beavers or bat-men. It matters not only for this particular story, but also for the ways in which we think about public debate in relation to expertise. Visible to all, especially as public spectacles,

Frontispiece of Hanes y Lleuad, the Welsh translation of the Moon hoax
Hanes y lleuad: yn gosod allan y rhyfeddodau a ddarganfyddwyd gan Syr John Herschel, trwy gynnorthwy gwdyr-ddrych …
(An abridged version of R. A. Locke's 'Some Account of the great astronomical Discoveries lately made by Sir John Herschel'), c. 1858

the reporting of scientific discoveries has been critical to the creation of 'news' as a specific kind of objective commodity. The reception of Locke's satire highlights the way in which public judgement in democratic societies depends on the accessibility of information and the ability of readers to discriminate fact from fiction.

At least since the emergence of cheap papers such as the *Sun* nearly two centuries ago, science has been a resource for ideals of objectivity, but it has never been a stable one. Scientific experts depend on the reading public both as potential contributors and as a repository of ideas about how science should work. Their labours are not separate from the journalistic reports, but are bound up with them in a communication cycle. Science can never serve as the foundation for the creation of a narrowly empirical, fact-based version of objectivity in the mass media; rather, ideals of objectivity depend upon the integrity of public debate. There was, the Moon story reminds us, no escape from democratic judgement, open channels of communication and a critical, informed public in determining whether or not published reports are true.

Nouvelles découvertes dans la Lune Faites par Sir John Herschel dans son observatoire du cap de bonne espérance …
(New discoveries on the Moon made by Sir John Herschel in his Cape observatory)
Thierry Frères, c. 1835–38

Fly Me to the Moon: From Artistic Moonscape to Destination

Melanie Vandenbrouck

For as long as astronomers have been observing the Moon, astronomy, the visual arts and popular culture have been intricately linked. Galileo's telescopic observations and early works of science fiction captured people's imaginations through the power of images and of words. While astronomers may have begun the depiction of the Moon as a world, the visual arts and popular culture would inspire and be inspired by technological creativity as humankind reached closer to the Moon, leading to the epoch-making 'small step' in 1969. In 1978, a Chinese poster inviting the new generation to reach for a Moon inhabited by the goddess Chang'e showed how the idea of lunar travel remained, well into the twentieth century, equally rooted in folklore and progress, age-old aspirations and contemporary political context, each reflected in the arts. In the representation of the Moon and lunar travel, the blurring of the lines between fantasy and fact leaves a space for the imagination to run free and lead progress.

When in 1766–72 Fillipo Morghen produced *Raccolta delle ... famose viaggio dalla terra alla luna* (A Tale of the Famous Travel from the Earth to the Moon), he imagined a fertile lunar world. Its tiny inhabitants lived in tomato houseboats and tree-houses, travelled on pumpkin-shaped vessels or domesticated butterfly-winged snakes, fought giant rats and kept snails in cages. Morghen's lunar fantasy displayed the Moon's otherworldliness and its inhabitants' otherness in the visual language of his time: a rococo style replete with chinoiserie ornamentation, New World figures, an exotic landscape of marshlands and floating habitations. The series of etchings was dedicated to John Wilkins, whose *Discovery of a World in the Moone* (1638) claimed that the Moon was inhabited and that one could go there. Fantastic literature and satirical prints were by then awash with weird and marvellous ways to travel to the Moon, from being drawn by a flock of geese, to solar- or dew-powered contraptions.

At the end of the eighteenth century, John Russell had represented the Moon as a geological world of craters and plains and highlands, while his friend the astronomer Sir William Herschel had seen 'growing substances', forests and caverns in its womb and imagined 'inhabitants there of some kind or other'.[1] Some nineteenth-century artists continued to illustrate lunar travel in a rather romantic vein – Gustave Doré's 1877 illustrations to Ariosto's sixteenth-century *Orlando Furioso* being a case in point – and the 'Moon Hoax' had publicised a lunar arcadia not unlike Morghen's. But there were also serious attempts to imagine what this world, brought closer and into sharper focus by telescopic and photographic technologies, would look like and, by extension, how to reach it. Throughout the second half of the nineteenth century, moonscapes not only became more tangible, but also permeated visual culture so much that to reach for the Moon became the obvious next step. By the end of the century, astronomers had a clearer idea of its make-up, inspiring novelists and illustrators to imagine the Moon as a world and as a destination, often incorporating the technological developments of the time. To fly to the Moon was no longer a far-fetched dream. Artists made it believable.

As the artist and writer Ron Miller suggests, 'space art' is rooted in the aesthetic of the 'sublime', the apocalyptic visions of a John Martin in England or a Doré in France, and the landscape painters of the Hudson River School, who exposed the last frontier of the American wilderness. In his *Philosophical Enquiry into the Origins of our Ideas of the Sublime and the Beautiful* (1757), Edmund Burke (1729–97) theorised the sublime as the awe-inspiring spectacle of nature's vastness, greatness and intensity, eliciting astonishment, horror and, secure in the knowledge of one's relative safety, pleasurable terror. Burke's sublime is given perfect representation in Irish theologian and astronomer Josiah Crampton's *The Lunar World* (first published in 1853). Early in his book, Crampton invites the reader to:

> *stand ... within the arena of Tycho. Around us, on every side, stands a mighty wall of rock, extending in a circle of 150 miles, or 54 in diameter. Looking up from the interior plain, it is 17,000 clear feet of precipice before the eye rests. Before us extends a*

Jin Dingsheng, The Youth of the New China Should Fly 飞吧，新中国的少年，
published by Jiangsu People's Publishing House, May 1978

Fillipo Morghen, Raccolta delle … famose viaggio dalla terra alla luna, *(A Tale of the Famous Travel from the Earth to the Moon)*

'Lunar crater', from The Lunar World: its Scenery, Motions, etc (Considered with a View to Design), *by Josiah Crampton*

> *plain for about 25 miles, interrupted, however, by concentric circles of rocky mountains, or barriers, that encircle (in irregular and broken masses, of fearful magnitude and height) the awful centre, whence, from a black and profound gulf, that opens its mighty jaws, springs a huge, dark mountain […] whose steep and pointed summit, higher than the lofty Snowdon, shoots upward for above 4,000 feet in sheer precipice from the plain.*[2]

Unlike earlier illustrated accounts that presented fantasy lunar worlds, Crampton's included the first depictions of moonscapes that aspired to geographical verisimilitude, albeit within the sublime aesthetic, which found perfect expression in this alien world chiselled by a violent history of impacts and the interplay of light and shadow. The verticality, magnitude, irregularity, might and darkness of this 'terrible' lunarscape read like a textbook account of Burke's sublimity. 'The awful character of such scenery is feebly represented by anything terrestrial analogous to it', Crampton continued.[3] Analogies with earthly features, however feeble Crampton might feel them to be, allowed for the reader/observer to make sense of the Moon. So too had the nomenclature devised centuries earlier, which described the Moon's features as seas, ridges and valleys. Now these features were given recognisable shapes.

Although Crampton's book is seldom mentioned in selenographic studies, it was admired by James Nasmyth, co-author with James Carpenter of the celebrated *The Moon: Considered as a Planet, a World, and a Satellite* (first published 1874), who felt an affinity with his predecessor's project. The archetypal Victorian polymath, Nasmyth had artistic talents as well as a being an inventor, engineer, philosopher, amateur astronomer and photographer. The book's illustrations were the result of years of painstaking observation of the Moon, whose features Nasmyth delineated in sketches and paintings in the 1840s and 50s, before giving them form in plaster reliefs that were photographed for the book. In addition to his magnificent Full Moon (plate IV), the book includes telescopic views of specific features, as well as imaginary 'landed views', that is, visualisations of what the Moon would look like as seen from its surface. The son of Scottish landscape painter Alexander Nasmyth, James Nasmyth understood that the Moon would look all the more tangible if grounded in familiar features. His lunar scenery thus uses the visual language of the time, although Nasmyth recognised the impossibility of conjuring up a world on which human eyes had never gazed. A picture that 'cannot but fall very, very far short of the reality', *Aspect of an Eclipse of the Sun by the Earth as it Would Appear as Seen from the Moon* (plate XXIV) shows a rocky range plunged into crimson darkness as the eclipse reaches totality, the surreal colours caused by the effects of the absorption of the Sun's light by the 'vapours of our atmosphere'. The astronomical event climaxes with 'the grand earth-ball, hanging in the lunar sky, like a dark sphere in a circle of glittering gold and rubies'.

In the equally dramatic *Group of Lunar Mountains, Ideal Lunar Landscape* (plate XIII), the Moon presents jagged mountains of awesome height, their peaks carved by the Sun's uninterrupted brightness. They 'seem to float like islands of light in a sea of gloom. […] the crater rim glistens like a silver-margined abyss of darkness'. Nasmyth resorted to photography to render the otherworldly angle and length of the lunar shadows, staging three-dimensional plaster reliefs under the oblique light, as in *Plato*'s elongated, black shadows (plate XIV). Nasmyth predicted that the 'violent contrast' of brightness and darkness, and the 'awful blackness' of the shadows themselves would be unnerving to a lunar spectator, 'in every sense unearthly and truly terrible'. In this hostile world:

> *the absence of all the conditions that give tenderness to an earthly landscape; […] the entire absence of vestiges of any life save that of the long since expired volcanoes […] make up a scene of dreary, desolate grandeur that is scarcely conceivable by an earthly habitant.*[4]

Nasmyth and Carpenter's book presented cutting-edge astronomy. Its illustrations of the Moon's atmosphere-less sky, intense sunlight, deep chasms and volcanic mountains also proved influential. The eminent French astronomer and science fiction writer Camille Flammarion's 1877 *Les Terres du Ciel* (Heavenly Worlds), containing diagrams, maps and other scientific visualisations (including reproductions from Nasmyth and Carpenter), was expanded in 1884 to include evocative full-plate illustrations, inviting the armchair traveller on a celestial journey across the worlds of the Solar System. While it exposes the latest knowledge and expounds on folklore and myths, the text is written with poetic flourish. Some of the whimsical images may reflect Flammarion's unconventional ideas about extra-terrestrial life and cosmic reincarnation but also echo contemporary

'Aspect of an Eclipse of the Sun by the Earth…', from The Moon: Considered as a Planet, a World, and a Satellite *by James Nasmyth and James Carpenter*

'Plato', from The Moon: Considered as a Planet, a World, and a Satellite *by James Nasmyth and James Carpenter, detail of the original plate*

art of the time – the plates by the illustrator Motty are symbolist in feel. Most, however, are scientifically up-to-date, credible visualisations of the book's content. The chapters on the Moon include illustrations that had become familiar tropes of lunar imagery. Paul Fouché's Sun eclipsed by the Earth and brightly illuminated craggy peaks are both in Nasmyth's line.[5] His *Phases de la Terre vue de la Lune* (Earth's phases seen from the Moon),[6] showing Earth's crescent against a star-studded sky above a mountain range, would be followed by many such views (pre-empting Apollo 8's 1968 *Earthrise*). Common to these, of course, is a reassuring familiarity: presenting the Moon as a mirror to our world, its landscapes and vistas resembling those on Earth – otherworldly and familiar, tantalisingly close yet still out of reach. Half a century later, Lucien Rudaux's views of Earth seen from the Moon and broodily dark moonscapes in *Sur les autres mondes* (On Other Worlds, 1937) prefigure the revelations of images produced during the Space Race. Rudaux's lunar hills are noticeably more rounded, with softer edges than those of his predecessors. An astute observer, the commercial illustrator, amateur astronomer, and first artist to specialise in astronomical art, Rudaux had correctly deduced that the Moon's sharp shadows were not a reflection of its terrain, but of the oblique light hitting its surface. This did not preclude truly dramatic illustrations, such as *Spectacle au fond d'une grande crevasse aux parois abruptes* (Spectacle in the Depth of a Large, Steep Crevasse), which seems to prefigure scenes from Stanley Kubrick's *2001: A Space Odyssey* (1968), magnificent sunrises and an evocative night scene bathed in earthlight.[7]

Parallel to the publication of popular and increasingly affordable illustrated astronomical books aimed at an ever wider public, the nascent art of cinema proved influential in fuelling the public's desire to reach the Moon. Indeed, as Richard Dunn explains in his essay earlier in this section, real space technology and that fictionalised in popular culture continuously bounced off each other in the decades leading up to humankind's first steps on the Moon. While the technology showcased in motion pictures became increasingly authentic, the moonscapes in the earliest films did not. To depict a moonscape devoid of human presence is not the same as staging entertainment-filled exploration, and plausibility could be sacrificed to dramatic effect. Shot in the *féerie* stage tradition of fantasy plots and spectacular set design, George Méliès's *Voyage dans la Lune* (A Trip to the Moon, 1902) presented the lunar world as a jumble of geological features, with theatrical snowfall, giant mushrooms and Selenites dressed in tights. While Fritz Lang's ground-breaking *Frau im Mond* (Woman in the Moon, 1929) was lauded for the accuracy of rocket technology and the effects of space travel (pet mouse Josephine's floating cage makes for a great demonstration of zero gravity), the far side of the Moon concealed caves full of gold, had a breathable atmosphere and looked like a skiing resort in the Alps *sans chalets*.

Visual arts and film would play an important role in the run up to Apollo. A key figure was Chesley Bonestell (1888–1986). Architect by profession and training, amateur astronomer and space artist by passion, Bonestell reinvented himself in his late 50s by publishing visualisations of the Solar System in the magazine *Life* from 1944. This led to an introduction to popular science writer and *Frau im Mond* advisor Willy Ley (who had fled Nazi Germany in 1935), with whom he collaborated on publications that would install Bonestell as the father of space art and godfather of the American space programme. Among these, The *Conquest of Space* (1949), in which he illustrated Ley's text with 58 paintings, would become a popular science classic. Miller credits it with 'convinc[ing] an entire generation of Post-World War II readers that spaceflight was possible in their lifetime'.[8] Fledgling science-fiction writer Arthur C. Clarke's review of this 'outstanding example of cooperation between art and technology' in *Aeroplane* indicates how influential Bonestell would become:

> *Mr. Bonestell's remarkable technique produces an effect of realism so striking that his paintings have sometimes been mistaken for actual colour photographs by those slightly unacquainted with the present status of interplanetary flight ... To many, the book will for the first time make other planets real places, and not mere abstractions.*[9]

This photorealism was key to Bonestell's success and that of the projects on which he worked.

Around the time he turned to astronomical illustration, Bonestell became a successful visual effects artist for Hollywood films. In 1950, he contributed to George Pal's and Irving Pichel's *Destination Moon*, which shared much with *Frau im Mond*, from the plot's motivation driven by commercial profit, and the sub-plot of self-sacrifice and sang-froid, to the meticulously detailed technology of the *Luna* spaceship, shaped like a V-2 rocket. This time, though, the setting was to be given as much scientific credence as the

Friede on the Moon, still from Fritz Lang's Frau im Mond, *1929*

rocketry. Bonestell provided the matte paintings that form the movie's backdrops, notably the 14 foot (4.27 m) panoramic moonscape explored by *Luna's* crew. Robert A. Heinlein, technical adviser and author of *Rocket Ship Galileo* on which the screenplay is partially based, later recalled:

> *I had selected the crater Aristarchus [for the lunar landscape]. Chesley Bonestell did not like Aristarchus; it did not have the shape he wanted, nor the height he wanted, nor the distance to an apparent horizon. Mr. Bonestell knows more about the surface of the moon than any other living man; he searched around and found one he liked – the crater Harpalus, in high northern altitude, facing the Earth. High altitude was necessary so that the Earth would appear near the horizon where the camera could see it and still pick up some lunar landscape. Northern latitude was preferred so that the Earth would appear conventional and recognizable.*[10]

And yet, Bonestell's lunar soil, scarred by mud cracks, might be effective in suggesting a barren world, but its anachronism on the waterless Moon betrays artistic licence even more exaggerated than the craggy peaks he favoured.

The film was not just prescient in terms of technology. Taking his first steps on the Moon, mission lead Dr Cargraves proclaims: 'in the name of the United States of America, I take possession of this planet on behalf of and for the benefit of all mankind', pre-empting by 19 years Apollo 11's commemorative plaque stating that 'we came in peace for all mankind' (which, while prevented by the Outer Space Treaty from making an overt statement of sovereignty, at least decreed the USA as representatives of humanity). Cargraves then transmits his first impressions to Earth: 'utter barrenness and desolation ... the sky is black, velvet black', words echoed in Buzz Aldrin's 'magnificent desolation' in 1969, and his fellow astronauts' visions of darkness. *Destination Moon* was well received at the time, leading *New York Times* critic Bosley Crowther to write: 'they make a lunar expedition a most intriguing and picturesque event', no doubt thanks to Bonestell's awe-inspiring moonscape.[11] As Dunn shows in this volume, the film's closing credits heralded the dawn of the space age, as the USA was to unwittingly embark on its most expensive national endeavour in 1955 after announcing plans to launch an artificial satellite into space, triggering the Soviet Union's launch of *Sputnik 1* in 1957 and the well-known chain of events afterwards.

The popularisation of space exploration loomed large in Fifties America, and in addition to popular books and films, from 1952 to 1954, Bonestell contributed to a series of articles written by Wernher von Braun and Ley for *Collier's* weekly illustrated journal. One of America's most popular magazines, *Collier's* included news, photojournalism, short stories and serialised novels. Together with illustrators

The Conquest of Space, *by Willy Ley and Chesley Bonestell, 1950*

Pressbook for Irvin Pichel and George Pal's Destination Moon, *1950*

Fred Freeman and Rolf Klep, whose efforts he coordinated, Bonestell gave form to Ley's and von Braun's vision of future space exploration. Bonestell reflected the sentiment of many in enlisting von Braun's expertise, when, upon being 'chided for choosing a "Prussian Nazi instead of an American engineer"' he reportedly retorted: 'Because he had more successful experience building rockets, and more faith in going to space than anyone else than I could find'.[12] The ends of conquering space justified the means. (Indeed, having been the leading V-2 engineer during the Second World War, von Braun had surrendered to US forces at the end of the war, and eventually became a key figure of the American space programme as the main architect of the Saturn V rocket used in the Apollo missions.) The articles predicted that the USA would launch a satellite into space by 1963, a crewed expedition to the Moon by 1964, before setting its sights on Mars. Though wildly ambitious, the articles insisted that the technology was there and that the costs would be reasonable. History would decide otherwise.

Nonetheless, the *Collier's* illustrations proved immensely influential to the era's obsession with space travel. As Miller puts it, 'If anyone had to illustrate a rocketship it had to look like a *Collier's* rocket or it just wasn't right. These were the standard'.[13] Bonestell's moonscapes would, of course, be proven 'wrong' by the images sent back by the Soviet and US missions. In 1956, the Boston Museum of Science had commissioned him to produce a 10 by 40-foot (3 x 12.2 m) panorama of the lunar surface for their Hayden Planetarium (completed in 1957, now at the Smithsonian National Air and Space Museum). After the Apollo 11 mission, the mural was swiftly rolled up and relegated to the Museum's stores. Earlier considered cutting-edge in its realism, it had been made obsolete by the revelation that the Moon's terrain was, as Rudaux had predicted, smooth rather than jagged, its hills more akin to sandy dunes than alpine peaks. Bonestell's reaction was telling: '[The Moon] looks for all the world like the Berkeley Hills'.[14]

While Nasmyth, Rudaux and, in their wake, Bonestell had endeavoured to represent the Moon as it would look to someone standing on its surface, it was a Soviet probe that would first show this in reality. With *Luna 2* as the first probe to impact on the Moon and *Luna 3* the first to send images of the far side (both 1959), *Luna 9* was another notable Soviet victory as the first man-made object to make a soft landing on the Moon on 3 February 1966. Its integrated television camera system undertook a photographic survey of its immediate environment, beaming eight hours of radio transmission and TV footage to Earth that, once stitched together, gave the first panoramic views of the lunar soil. (Ironically, the pictures were released worldwide after being intercepted by the UK's Jodrell Bank Observatory.)

To represent a land is to lay claim to it, and the moonscapes photographically captured by Soviet probes, and, later on, American astronauts, contributed to the discourse of dominance – technological as much as cultural, if not geopolitical – at play in the Space Race. In this context, one may read Vija Celmins's *Moon Drawing (Luna 9)* (1969) as a comment on the events taking place on and around the Moon during that period. Indeed, it is tempting to see Celmins (born 1938), who grew up in the USA but was born in Latvia shortly before the Second World War and her country's subsequent annexation by the USSR, as a particularly acute interpreter of these events. Celmins's work is distinguished by a photorealism achieved through painstakingly accurate reproduction, in drawing or painting, of photographic imagery. While her artworks are often long in the making, with their deliberate slowness bearing meditative qualities, it is unclear whether she started the drawing in 1966, and when exactly in 1969 it might have been finished – that is, before or after America achieved its momentous 'first' with Neil Armstrong's 'small step'.

One can read the act of drawing a piece that took months if not years to complete as a parallel to what was unfolding between the USA and USSR, the intense scrutiny to which the lunar surface was subjected by both nations thus matched by the artist's own. As Celmins confesses, while her process might seem 'tedious for some',

> *[...] for me it's kind of like being there. I've always liked scientific images because it's kind of anonymous and often the artist for the image has been a machine. I like the idea that I can relive that image and put it in kind of human context, and I like you to be able to scrutinise it and sort of relive the making of it the way that I have been doing.*[15]

It may be that when she started drawing, access to the Moon was still mediated by machine; even at the time of the Moon landing, it was mediated by the 'moonwalkers' and their Hasselblad cameras. In reinterpreting *Luna 9's* image, Celmins, and the spectator with her, are performing a vicarious journey to the Moon.

The rivalry between the two superpowers was played out in the arts too. The superiority of the Soviet space programme and its achievements were charted in propaganda posters throughout the Sixties. In a heavily symbolic visual language, these championed collective effort to ensure the triumph of the nation and proclaimed the glory of the Communist Party. Hailing heroic citizens advancing the common cause of conquering space, they also celebrated canine cosmonauts, Belka and Strelka (1960), the first man in space Yuri Gagarin (1961) and the first woman Valentina Tereshkova (1963). One poster, its title translated as *Peace for the People* (1964), (see page 162) evokes the fact that, despite the Space Race being one of the Cold War's most openly fought contests, both sides proclaimed their lunar ambitions to be emblematic of a new era of peace and technological prosperity. In 1963, President Kennedy had even suggested that the Moon be a joint endeavour between the two, but was rebuked by Soviet leader Nikita Khrushchev. *Peace for the People* shows the USSR celebrating another first, the launch of a three-man spacecraft, *Voskhod 1*, in October 1964. The soaring rocket expresses the USSR's irresistible advance, the three cosmonauts steadfastly looking up as they reach for the stars.

In America, the response to the space programme was ebullient, not least within the circle of Pop artists. In response to the historic Moon landing, W.H. Auden had written in a poem of the same name:

> *It's natural the Boys should whoop it up for*
> *so huge a phallic triumph, an adventure*
> *it would not have occurred to women*
> *to think worth while ...*[16]

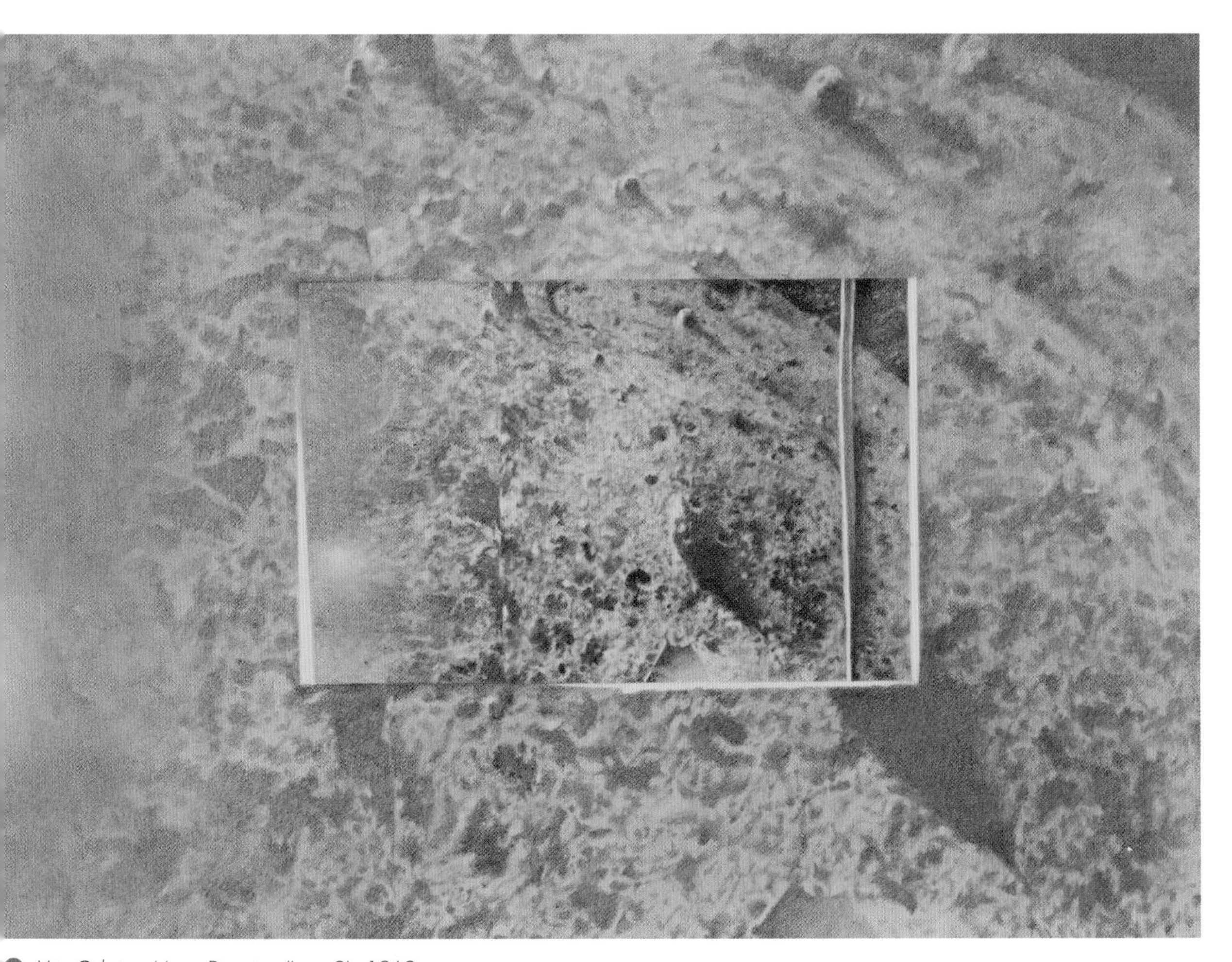

Vija Celmins, Moon Drawing (Luna 9), *1969*

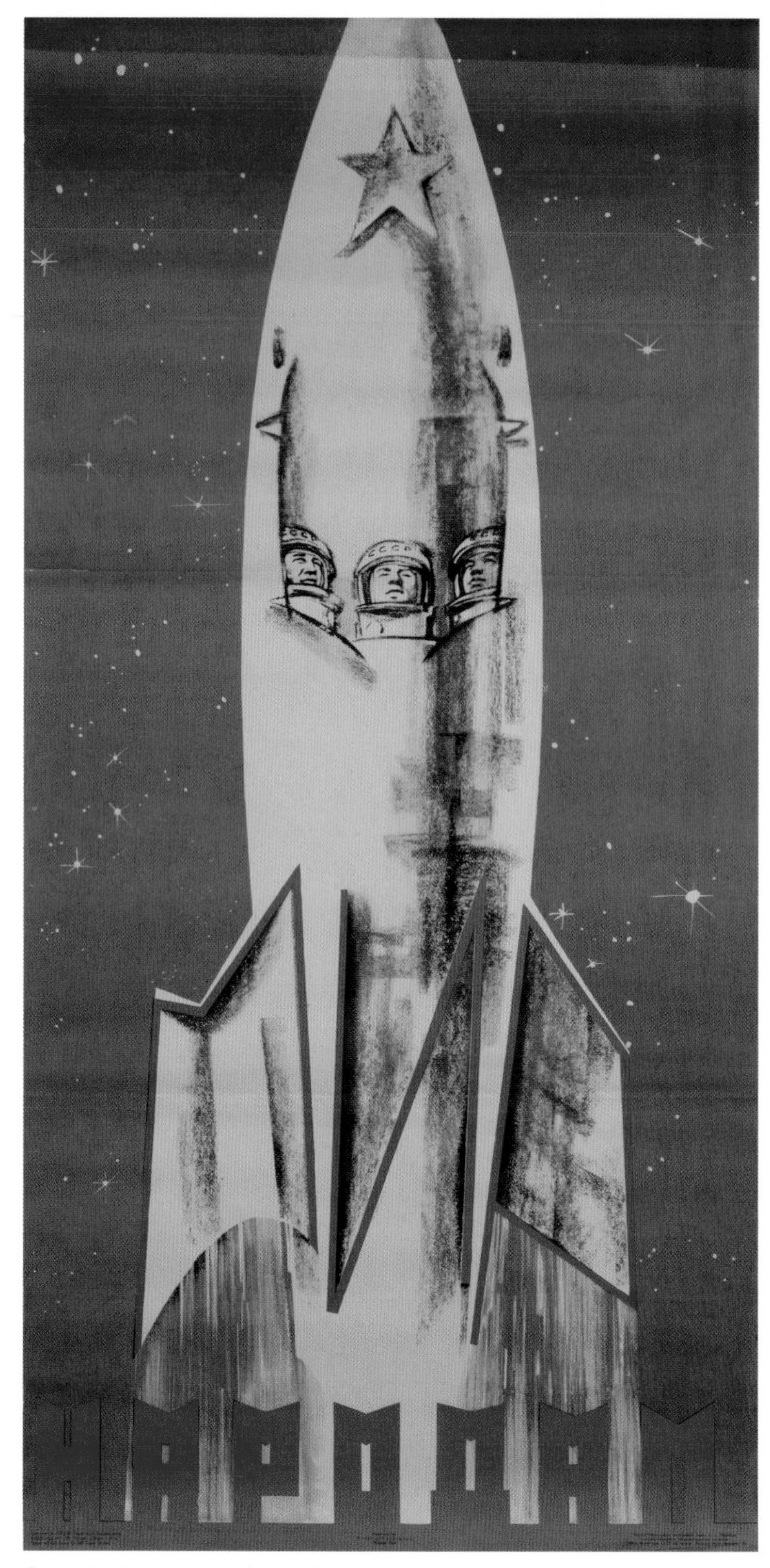

M. Gordon, Peace for the People, *1964*

In what could be seen as a misguided ode to womanhood, Auden not only underestimated women's involvement in the Space Race – from Tereshkova's space journey to the American space programme's thousands of women working behind the scenes as seamstresses, computer scientists and astronaut trainees – but also failed to register women artists' responses.

Like Celmins, Kiki Kogelnik (1935–97) turned her sight to the Moon in the Sixties. The Austrian artist had joined the Pop Art scene in New York in 1961 to embark on what curator Ciara Moloney calls 'her persistent representation of the delights and terrors of space travel'.[17] The Space Race could not have found a more attentive observer: 'I'm involved in the technical beauty of rockets, people flying in space, and people becoming robots', Kogelnik said.[18] *Fly Me to the Moon* (1963) (see page 164) presents her signature stencilled, featureless silhouettes, thrust into an upward motion, their vivid colours standing out on a background of Moon-grey dots, while a rocket heads downwards. There is an undeniable sexiness to this painting – are they locking fingers, or freeing themselves from their embrace, in ecstatic climax or release? – evoked further by the title that would have conjured up one of the catchiest tunes of the time to any New Yorker. Bart Howard's 1954 song 'In Other Words' had been retitled 'Fly Me to the Moon' in 1963, the year in which Kogelnik produced her painting, and was sung by the likes of Peggy Lee, Nat King Cole, Sarah Vaughan and Brenda Lee. Frank Sinatra's 1964 version would become associated with the Apollo era and was the first music to be heard on the Moon. By then, the US space programme was still lagging painfully behind the USSR's but Kennedy's 'We choose to go to the Moon' speech, delivered on 12 September 1962, instilled a sense of optimism and progress, while Tereshkova's achievement no doubt resonated powerfully for Kogelnik.

What was it like to float in space, to embark on uncharted journeys, to step on another world; Kogelnik's work seemed to be asking. Experimenting with the body and the moving image, her *Untitled (Floating)* (c.1964) expresses the exhilaration and disorientation of zero gravity. The effect is simplistically achieved – the camera turned upside down and tilted, Kogelnik standing in the corner of a room, waving her arms and swaying her torso – but the makeshift quality of this work, at odds with the cutting-edge technology being developed by NASA, is perhaps what makes it so subversive.

This focus on the lived experience of being there – 'here and now' – not just witnessing, but also participating in the unfolding of humankind's attempts to reach another world, found further expression in the *Moonhappening* Kogelnik staged in Vienna as the world's eyes and ears were locked on the Apollo 11 crew's first moonwalk. Kogelnik issued silkscreen prints of the Full Moon – a familiar sight her contemporaries would never see the same way again – on which she wrote the first impressions of Armstrong and Aldrin in real time as they were broadcast to hundreds of millions across the globe. Kogelnik's space-age effervescence was only matched in intensity by the more sombre work she produced around then, featuring macabre forms, skulls and grinning faces, reflecting a growing unease about the effects of technology in the context of the Vietnam War, the civil rights movement in the USA and Cold War fears of nuclear annihilation.

In contrast to Kogelnik, Robert Rauschenberg (1925–2008) held an extraordinary position as an artist commissioned by NASA to chronicle the first crewed flight to the Moon. Invited to attend the Apollo 11 mission launch in Florida, Rauschenberg was granted access to Cape Kennedy's facilities, official photographs and technical documents, the astronauts and NASA personnel. The resulting series of 33 lithographs, *Stoned Moon*, expresses the sense of excitement and momentum – indeed, elation not unlike that provided by hallucinogenic drugs, as suggested by the word 'stoned' – that overtook him and the American public alike. Among these, the explosion of colours of *Sky Garden* evokes Rauschenberg's sense of awe and wonder upon witnessing the Saturn V lift off on 16 July 1969. With its distinct verticality and vibrant colours, this monumental lithograph conjures up the formidable blast and irresistible soaring of the most powerful rocket ever made, burning 770,000 litres of kerosene and 1.2 million litres of liquid oxygen as the ground trembled and the surrounding atmosphere sizzled. In the artist's own superlative words: 'POWER OVER POWER JOY PAIN ECSTASY, THERE WAS NO INSIDE, NO OUT, THEN BODILY TRANSCENDING A STATE OF ENERGY, APOLLO 11 WAS AIRBORNE'.[19] Superimposed with suggestions of the lush flora and fauna of Florida, the rocket's technical diagram proclaims Apollo 11's unprecedented scientific and technological achievement. *Brake* is more reflective, in understated monochrome black ink. It commemorates the loss of the Apollo 1 crew, Virgil 'Gus' Grissom, Edward White and Roger Chaffee, seen in the bottom right corner, who were killed in a fire in their capsule during a launch test in 1967. Like Belgian artist Paul van Hoeydonck's later *Fallen Astronaut* (1971), an aluminium figurine left on the Moon by the Apollo 15 crew as a permanent memorial to the 14 astronauts and cosmonauts who died during the Space Race, this rare consideration of the human cost complicates that period's overwhelming narrative of American 'manifest destiny', political, technological and imperial prowess.

While the *Stoned Moon* series evoked a sense of optimism, Rauschenberg had, like Kogelnik, become increasingly disillusioned by his country's involvement in the Vietnam War and the growing social unrest fuelled by poverty and racial discrimination. Following the NASA commission by a year, his screenprint *Signs* (1970) places the Apollo 11 achievements within the wider, fraught context of Sixties' America: the collage includes Aldrin and Armstrong alongside GIs in Vietnam, the

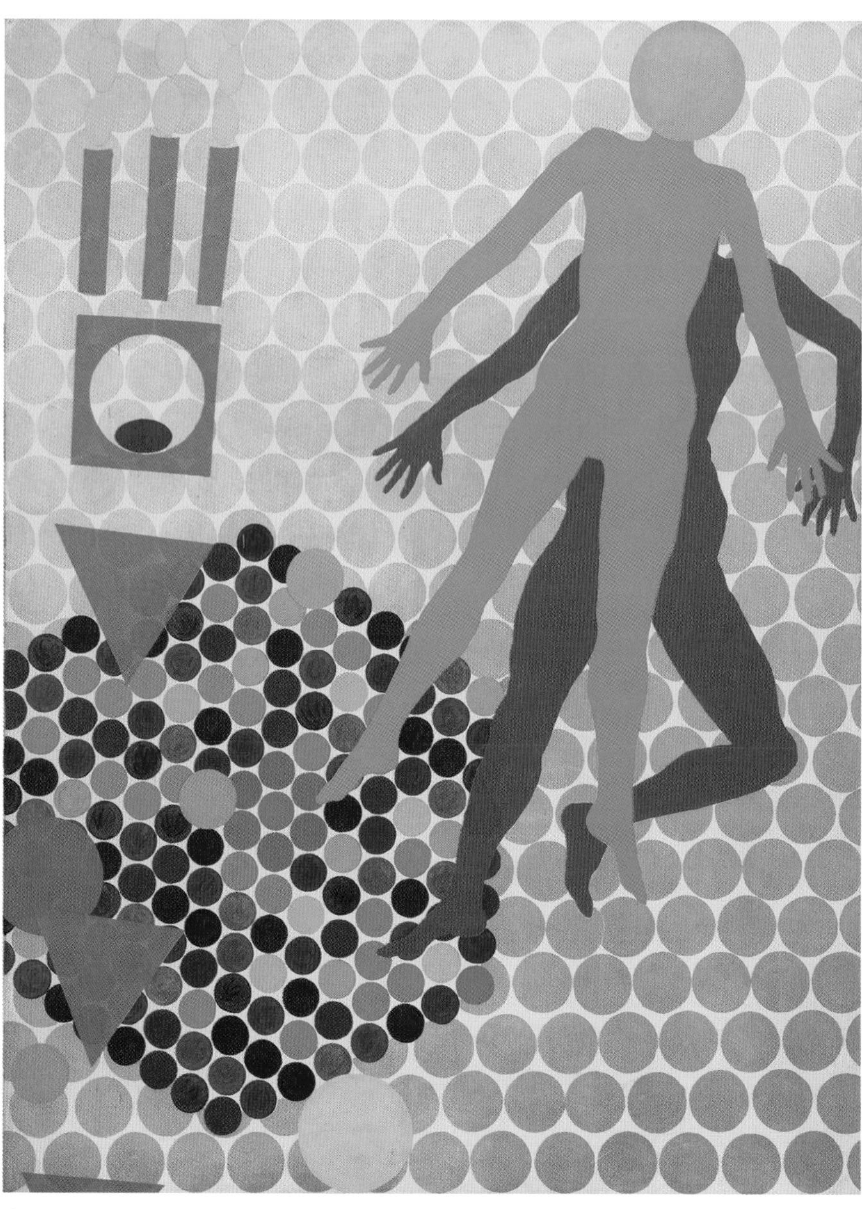

Kiki Kogelnik, Fly Me to the Moon, *1963*

Robert Rauschenberg, Brake, a lithograph print from the Stoned Moon series, 1969

bodies of Martin Luther King (murdered in 1968), John Fitzgerald and Robert Kennedy (assassinated in 1963 and 1968 respectively), peace protesters, the bloodied figure of a man wounded in a race riot and anti-war singer Janis Joplin.

More enigmatically, Mark Rothko's final series, *Untitled (Brown on Gray)* (1969–70), irresistibly recalls a moonscape even as it eschews interpretation. In each work in the series, Rothko (1903–70) reduced his palette and composition to the opposition of a dark brown, almost (and sometimes) black, and a lighter, luminous grey. As art historian Bryony Fer puts it, 'If there is nothing much there, it is an extraordinarily palpable void', words that could equally be applied to a lunar landscape, echoing the Apollo astronauts' accounts of the textured, velvety darkness they witnessed.[20] Rothko emphatically rejected figurative or even symbolic readings of his 'relentlessly abstract' art.[21] But contemplating the upper register, inscrutable like a starless lunar sky, and the soft dusty grey of the lower field's open brushwork, separated by an ever-so-subtle concave horizon line, one cannot help but wonder whether this epoch-defining moment might have found itself on Rothko's canvas, deliberately or not. An intensely private pursuit that did not, unlike previous work, rely on a commission and its public display, the series was solely to occupy the last few months of his life, from the summer of 1969 to early 1970. It is not inconceivable that the all-pervasive Apollo frenzy permeated the walls of Rothko's studio – he was, after all, an avowed TV enthusiast.

The Russians may have claimed many milestones in the Space Race, but it was the Americans who became the uncontested winners by landing human beings on another planetary body. British Pop artist Peter Phillips evokes this in *Apollo II* (1969) with its American bald eagle adorned with Apollo 11 mission patches, tightening its claws above the lunar surface.

And yet, was the achievement of that long-held dream just smoke and mirrors in the face of US domestic and global difficulties? A few months before his death in 1987, Andy Warhol (1928–87) published an edition of prints that conveyed, more powerfully than any other work from the Pop artist, the lasting impression the American space programme had etched on his memory. *Moonwalk* takes on two iconic NASA photographs taken by Neil Armstrong during the historic Moon landing: Buzz Aldrin standing in profile near the American Stars and Stripes; and the face-on shot in which Armstrong's silhouette is reflected in his companion's gold-plated visor (the image Rauschenberg had used for *Signs*). Merging them into one scene in vivid, Pop Art tones, *Moonwalk* questions the accuracy of our memories of the Moon landing, while also conjuring a sense of lost euphoria and nostalgia-tinged glamour.

By 1987, Apollo had become a fading memory in the American psyche. The Moon, no longer the 'next frontier' it had been in the Sixties, was but a distant dream. It would once have seemed unthinkable to an artist who claimed that he painted the walls of his Factory silver because of 'space, the astronauts', that the programme should end so abruptly and with so little fanfare in 1972. In 1969, Warhol was one of six artists laying claim to their bit of the Moon, by arranging with the crew of Apollo 12 that their thumbnail-sized *Moon Museum* (1969) would be the first artefact, other than flags and Space Race litter, to be left on its barren plains. Warhol's contribution was unambiguously provocative – part rocket, part penis, part his own initials. Claes Oldenburg contributed an equally tongue-in-cheek yet quintessentially American Mickey Mouse, Rauschenberg a single line, suggesting, perhaps, that straight lines, not found in nature, are artificial, man-made constructs. Whether they were questioning the hubris of the American programme, embracing it, or neither, the Moon, they seemed to say, belonged to artists.

Peter Phillips, Apollo II, *screenprint, 1969*

Fashion and the Moon Landings

Susanna Cordner

In the October 1967 issue of *American Vogue*, in response to the US government's endeavours to 'put a man on the Moon before the end of the decade', the playwright Tom Stoppard described the disconnect he felt between the Moon as a literal landscape and its role as a cultural canvas:

> *You can't just land on the moon. It's much more than a location, it's a whole heritage of associations, poetic and religious. There are probably quite a few people around who'll go mad when the first man starts chumping around this symbol in size-ten boots.*[1]

Stoppard framed both the excitement and the anxiety surrounding the Apollo space programme: that it would allow humankind to reach new heights and spur us into a space age, but with it perhaps put an end to the Moon's symbolic role as a blank canvas on which to project your dreams and predictions for the future. The fact that Stoppard made this observation in a leading fashion magazine is fitting, given fashion's own dual relationship with the Moon as a source of inspiration and a projection space for futuristic fantasy. In fact, fashion's connection to the Moon landings is both direct – through the development of the spacesuit – and figurative. In both cases, fashion's interest in the Space Race lay as much in the journeying as in the destination. Perhaps predictably, this preoccupation was in part driven by what would, or could, be worn to get there.

Firstly, fashions of the Fifties and the spacesuit of the Sixties share a common foundation. After failed attempts by major military contractors and NASA engineers to design a successful spacesuit, the International Latex Corporation (ILC), better known by the name of their offshoot consumer company Playtex, became the prime contractor for the Apollo suit in 1965. As Playtex were, and still are, primarily purveyors of underwear, the race to space was therefore clothed and, in part, paved by a fashion company. In his brilliant book *Spacesuit: Fashioning Apollo*, Nicholas de Monchaux describes the ways in which, when developing the ILC A7L suit, the company utilised not only Playtex's technological and material expertise but their skills with a needle. When they landed on the Moon in 1969, Neil Armstrong and Buzz Aldrin were wearing custom-sewn spacesuits for which 'underwear seamstresses [had] to deploy couture techniques to fuse disparate layers at 64 stitches to the inch'.[2]

ILC's experiments were made possible by the demand for Playtex's rubber girdles. Playtex was indebted for the success of its product to French Haute Couture designer Christian Dior's New Look, a style and trend which, in 1947, reignited the fashion industry and centred on an extreme silhouette that overemphasised the female form. Underneath their dresses, followers of the New Look wore foundations that built up or pulled in their frames at the relevant points. The rubber girdle, compressing the waist and smoothing the tummy, was an essential component of this hidden architecture, and Playtex was the largest manufacturer of these garments in post-war America. Their commercial success in turn funded Playtex's pursuits and research in other areas, primarily their development of the spacesuit. This chain of commerce makes unlikely bedfellows of three design icons of the twentieth century. As described by de Monchaux, 'each, in its own way, acted as the literal foundation to a visual icon – whether the jet-age, New Look female of the Fifties or the space-age astronaut male of the Sixties'.[3]

The Moon missions became a design emblem and opportunity for designers to take a stylistic leap into the unknown, with many creating outfits inspired by what they imagined life would be like either on the spacecraft or on the Moon's surface. In the Sixties, the Moon provided a heady and uncanny mix of familiarity and otherness – a subject heavy with heritage and symbolism that was, simultaneously, still essentially unknowable, but not for much longer. With that sense of impending discovery, 'Moon-rush' fashions, as the trend was referred to by the press, also gave designers scope to look to the future. To associate your designs with the Space Race was to tap into a contemporary cultural and cosmic touchstone while also staking your claim to be a part of styling the future. As a result, while fashion forces were at work behind the seams of the spacesuit, key mid-twentieth century fashion designers such as André Courrèges, Pierre Cardin and Paco Rabanne

ILC Group Leader and seamstress Hazel Fellows, at work on a Big Moe sewing machine. She is attaching the shell, liner and insulation layers of a Thermal Mircometeoroid Garment. Taken in 1967.

Andre Courreges, fashion designer, with his model in 1970

Family modelling space suits by Pierre Cardin, 28 April 1967

in France, were preoccupied with conjuring up their own predictions and interpretations.

With his 'Space Age' collection in 1964, André Courrèges launched what became known as his 'Moon Girl Look'. His white A-line mini dresses with contrast trims in vivid shades and silvers teamed with goggles and helmets inspired by spacesuits popularised Space Race fashion. Despite, or perhaps because of, these futuristic leanings, Courrèges was incredibly commercially successful. He sold over 200,000 skirts in 1965 alone and his designs, such as his square-toed white boots, were widely disseminated. His aesthetic remains a reference for designers such as Christian Lacroix today. Edging closer to the Apollo 11 mission, in 1967 Pierre Cardin released his 'Cosmos outfits' for men and women. These ensembles were also sometimes referred to as spacesuits by the press. Cardin's interest in the race to the Moon was primarily technological, with focus on the fibres and fabrics of the future and the scientific progress the Apollo missions marked. In 1968, he developed and released his own high-tech fabric, Cardine.

Both Courrèges and Cardin considered increased freedom of movement and practicality for their wearers as driving forces behind their space-worthy futuristic aesthetics. The success and enduring legacy of these designers perhaps highlights the difference between fashion's literal and symbolic relationships with the Moon. While they turned to our nearest cosmic companion for inspiration, they were under no pressure to create designs that would actually suit surviving on its surface, meaning that, while their work was stylistically 'other' to what had come before, it was still highly wearable.

The inspiration the beauty industry took from the Moon landings was more explicit. In 1969, advertising campaigns for make-up manufacturer Revlon's 'Moon Drops' range featured backdrops made of loose eye shadow heaped in mounds reminiscent of the cratered undulations of the Moon's surface. Available in iridescent colours, the products were designed to 'give your eyes the new 'luminesque look' – like great, gleaming opals held up to the light'.[4]

While the industry was at work mapping out their moves towards the Moon, individual style setters also celebrated the peak in progress embodied by the Apollo space programme and the cultural obsession with the lunar landscape. Professionally, Eddie Squires made his mark with iconic print designs, in particular 'Lunar Rocket', a screen-printed furnishing fabric that he designed for Warner & Sons Ltd (now Warner Fabrics), to commemorate the historic Moon landing. His personal fascination with this event is reflected and captured by a denim jacket by AMCO held in the V&A's collection. Bequeathed by Squires to the Museum, it was clearly a much-loved piece and project, which he customised and ornamented over years of wear. The sleeves and body of the jacket are covered in space-themed pins and patches, star studs and Apollo rocket prints teamed with repeats of appliqued pieces of his own 'Lunar Rocket' design. The jacket serves as a reminder of the ways in which references to the Moon landings acted as sartorial expressions and markers of both personal style and experience of an unprecedented event. In Squires's case, he wore his wonderment at the Moon on his sleeve.

Fifty years on, Stoppard's concern that landing on the Moon would stunt its role as metaphor and muse has proven unfounded, as contemporary fashion trends and references imply the reverse. The clearer the path to an understanding of the Moon has become, the more fashion has moved from imagining life on it, or the sartorial route to it, to indulging in the symbolism and superstitions that surround it. For both couture and cruise collections in 2017–18, Dior's creative director Maria Grazia Chiuri took inspiration from tarot cards, with the Moon as a repeated feature on accessories, embroideries and prints (see an example of a Dior scarf on page 243). In tarot terms, the Moon symbolises high emotion and intuition – sitting with uncertainty in order to let your dreams and unconscious guide you to a new path. In fashion terms, this symbolism may serve as an invitation to continue calling on the Moon to lead the way to new and innovative styles.

Eddie Squires's jacket, made by AMCO, c.1960–69, customised by Squires, c.1970–80

Space Cadets: Toys and Material Culture from *Sputnik 1* to Apollo 11

William Newton

The Space Race was one of the more intriguing proxies of the Cold War. Space travel moved from the realm of science fiction to that of science fact, and children's interests and material culture followed it. The breakneck progress made between the launch of *Sputnik 1* in 1957 and the first Moon landing in 1969 unfolded on television screens around the globe. Amazing first-time images of the sublime beauty of outer space, gleaming astronauts, and the long-held intrigue of the lunar surface appeared alongside a golden age of children's television. In parallel, the technological leaps that enabled the Space Race mirrored leaps forward in toy manufacturing technologies. During this period, space permeated all aspects of children's lives, from pocket money storage to bedspreads, toys to comic books. The combination of real and fictional TV programming and great toys captivated young imaginations worldwide.

Generally, children's material culture relating to the Moon and to space endorsed a spirit of optimism and faith in technological progress that permeated Western culture during the Fifties and Sixties. The style of toys produced around that period reflected the same ethos. This stemmed from a growing vocabulary of visual culture that was introduced through the Space Race. It was also a result of the sense of excitement caused by persistent advances, and from the fierce competition between the American and Soviet superpowers. Stated simply, at the start of the Space Race in the mid-Fifties, toys generally had a more fantastical, 'sci-fi' appearance. Once the Space Race took off, these toys gradually became more realistic, taking cues from actual missions, technologies and political ambitions in space.

After the Second World War, the Allies supported industrial recovery in Germany and Japan. Among the industries revived in both nations was toy-making, at which Germany had traditionally excelled. However, it was Japan that reaped the benefits of post-war investment, becoming the first Far Eastern centre for toy manufacturing. The pre-eminence of Japanese toys was established by the mid-Fifties, a result of direct financial support from the US, and preferential access to the American toy market. Crucially, Japanese manufacturers were able to perfect small battery-powered motors, which gave Japanese toys a superior range of movements and novelty features. Developments in lithographic printing processes allowed a finer level of detail to be applied to the surfaces of metal toys. Improved plastics technologies, developed during the Second World War, democratised access to toys by making them cheaper to make and buy. Plastics allowed an unprecedented level of detail, whilst also being modern, sanitary and in many ways 'space age'. These factors, along with the reduced costs of manufacturing, helped plastics take off as the most common materials for making toys.

When *Sputnik 1* was sent to orbit in October 1957, placing a human being in space, let alone sending one to the Moon, was still some years off. However, other milestones were exploited by manufacturers and the popularity of *Sputnik* proved to be a gift. A number of tin-plate 'manmade satellite' toys appeared that vaguely resembled the spherical form of *Sputnik*, but that were designed to be a good deal more exciting than the real thing. Some of these toys 'carried' animal crews of dogs, monkeys or bears, inspired by real-life counterparts such as Laika the Soviet dog, the first animal in space. These toys were brimming with character, and were an early indication of a change in the style of space toys, as they were explicitly influenced by real breakthroughs.

Spaceflight brought with it a new visual language of towering rockets, solar panels, and spacesuits in bright orange and shiny silver. These were quickly assimilated into toy design. At the start of the Space Race, there were numerous imaginative concepts for what spacecraft could be. Many were divorced from the reality of rocketry, instead resembling science fiction-style flying saucers. As the years passed and spaceflight developed, and human beings came ever closer to visiting the Moon, children came to expect greater levels of realism. The spacecraft and equipment used in real missions, and in activities such as spacewalking, began to be actively copied to make toys, collectibles,

Man Made Satellite, Yonezawa Toys, Japan, late 1950s

Capsule 7 mechanical toy, c.1965

model kits, confectionary packaging and posters. Now that science fiction had become reality, why pretend it looked very different to what could be seen on TV?

Astronauts and cosmonauts were pioneering heroes, some of whom circumvented political boundaries to become international celebrities. Importantly, being an astronaut was a career undertaken by real people, filled with excitement, discovery and new experiences. The tangibility and humanity of astronauts made them popular and set them apart from other types of childhood heroes, such as cowboys, pirates and knights. Their likenesses and stories were reproduced and made collectible in the form of stamps, books and cards. Spacesuits were highly distinctive and ultra-modern; they inspired fancy dress, both shop-bought and homemade. The BBC's long-running children's TV show, *Blue Peter*, broadcast demonstrations for how to make a papier-mâché space helmet in the run-up to the Moon Landing: any child could pretend to be a spaceman if they had some old newspapers.

Space was at first perceived as a boys' realm. In science fiction, women usually played the part of damsels in distress, or exotic love interests for red-blooded, heroic male space captains. However, in 1963 the USSR placed the first woman in space, Valentina Tereshkova, on what was only the twelfth crewed launch, creating an overnight inspiration for girls. US toy manufacturer Mattel responded two years later, in 1965, by producing a spacesuit for Barbie. Although falling short of sending a woman into space, making Barbie into a space explorer was more progressive than it might first appear. This Barbie presented being an astronaut as a potential career for girls, rather than just an opportunity for dressing-up. Her spacesuit was not wholly unrealistic as it bore a resemblance to early American models. As such, it could be said to have had more in common with action figures marketed at boys, such as GI Joe (Action Man), than with traditional fashion dolls.

From the moment President John F. Kennedy announced on 12 September 1962 that America would 'choose to go to the moon', putting boots onto the lunar surface became the Space Race's ultimate prize. What the moon would actually be like to walk on, or how humans might behave once they'd made it there, was open to speculation. Scientists did know that astronauts would probably move quite differently to how they did on Earth. This meant a brief but memorable revival of a novelty item called 'moon-shoes'. These were a popular outdoor toy comprising short,

Barbie as an astronaut, 1965

strong springs fixed to a footplate and worn like roller skates. The theory was that as a child hopped along on them, they would be able to experience (to a degree) some of the new sensations, such as low gravity, that were felt by astronauts in space.

As the Moon landing became inevitable, some manufacturers began to imagine how people might live there. One product of this was Tri-ang's much-loved *SpaceX* range, produced from January 1969. These colourful plastic vehicles were sold in groups of four, each representing a different zone: Earth Base, Space Station, Moon Base and Outer Space. The design of the vehicles was influenced by concepts produced by NASA, and from several film and comic book sources. The tone was one of optimism and intrepid exploration. Few could have predicted that interest in space would decline sharply following the real-life Moon landings, that there would be no real moon bases or space yachts.

In recent years, a renewed interest in lunar exploration has emerged at both geopolitical and domestic levels. The twenty-first century is an age of access in which people can connect to an abundance of information. Many children are now considered 'digital natives' to whom use of internet-enabled devices is second nature. A new golden age of television has materialised alongside unprecedented access to the 'retro' visual culture of the past. The narrative and immersive potential of video games is being realised, and in the past decade games have enjoyed a sharp upturn in serious critical respect. Access to all of these forms of entertainment has created a cocktail of content and aspiration, which stokes the renewal of interest in the Moon, science fiction and space. Parallels may be seen between now and the Space Race, when a technological present seemed to promise that living on a moon base would be only a few years away.

Moon shoes, made in England, 1950s–60s

Homemade astronaut's helmet made by Richard Blundel, Maidenhead, 1969

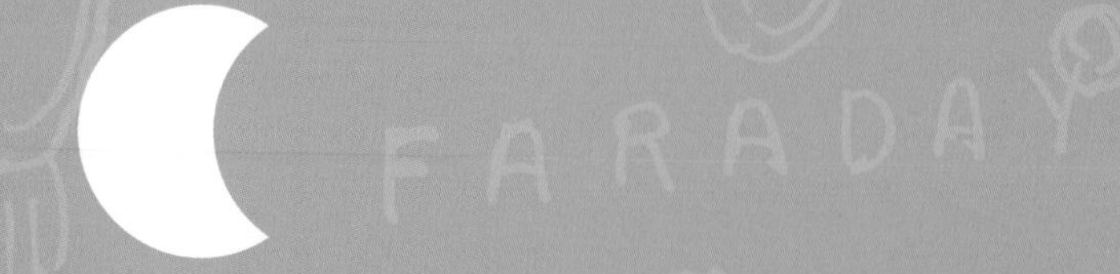

FOR ALL MANKIND?

When Apollo 17 blasted off the lunar surface in December 1972, astronaut Eugene Cernan became the last person to walk on the Moon. Humanity had achieved an age-old dream, but at what cost? Absorbing huge amounts of resources, Apollo had been largely motivated by the Cold War and many questioned its raison d'être. As the moondust settled, public and media interest in lunar exploration waned.

But, away from the limelight, the Moon's significance has continued unabated. The scientific data and samples returned by the Apollo missions have shed new light on the origins of the Solar System. The Apollo programme is also credited with inspiring a new generation of scientists and engineers. The Moon is now back on the scientific agenda as an object of study and plans are afoot to return with both robotic and human missions.

Many different players, including private businesses and new spacefaring nations such as China and India are now setting their sights on the Moon. As well as a base for research and a springboard to Mars, the Moon could be a source of mineral wealth or a platform for economic or military exploitation. Responding to these developments, artists are finding renewed inspiration in the Moon, questioning who it belongs to, interrogating future endeavours and exploring what our cosmic companion reflects back to us.

Melanie Vandenbrouck and Richard Dunn

The Scientists and the Moon

David A. Rothery

The existence and large size of the Earth's moon have been fortunate for the progress of science. Ancient Greek philosophers saw the curved edge of the shadow cast by the Earth onto the Moon during lunar eclipses as good evidence that the Earth is round. In the seventeenth century, Isaac Newton (1643–1747) proved the efficacy of his theory of gravity by demonstrating the relationship between the acceleration of objects falling at ground level and the motion of the Moon in its orbit. More recently, the Moon's proximity made it an achievable target that enticed humans beyond low-Earth orbit during the Space Race.

The Moon is not a planet in the astronomical sense: it is a satellite of Earth rather than of the Sun, but its size and composition mean that it has many similarities with the rocky planets Mercury, Venus, Earth and Mars. Studies of the Moon have advanced planetary science in general because geologists make no distinction between these five bodies when trying to understand how rocky worlds form and develop.

Craters

It was astronomers who first saw the Moon through telescopes, so they were the first people to be able to make scientific, evidence-based speculations about the nature of its surface. One of the most striking aspects of the Moon is that its surface is peppered with craters, as even a small telescope will reveal.

Robert Hooke (1635–1703, a contemporary and rival of Newton) used a 30-foot-long telescope to make a remarkably accurate drawing of a 200-kilometre wide region of the Moon where there is one large and several smaller craters (see opposite page). He then experimented to see how such a landscape could have been formed. When he boiled a mixture of powdered alabaster and water ('Plaster of Paris') he found that bursting steam bubbles produced convincing analogues to lunar craters. He also made plausible craters by dropping objects into wet clay, but rejected this as an explanation for the Moon's craters on the grounds that space was thought to be devoid of objects that could hit the Moon. Hooke therefore concluded that the Moon's craters were probably made by some kind of gas explosion. He was well aware that this is essentially what a volcanic explosion is on Earth, and drew the obvious analogy to volcanic craters.

Space probes have now provided close-up images of large and small bodies throughout the Solar System, including some relatively small moons of other planets, revealing that all ancient surfaces have craters of a vast range of sizes. It now seems obvious that craters must be made by impacts of cosmic debris. However, the volcanic hypothesis remained in vogue for most of the four centuries during which the Moon was the only body known to have abundant craters, although proponents of an impact origin were never quite silenced. The debate was enlivened from time to time by suggestions that would seem strange today; for instance, that its craters are the remains of lunar coral reefs.

Because the Earth lacks large, well-preserved impact craters to rival those on the Moon, geologists – wary of accepting exotic, unproven modes of origin – mostly favoured volcanic explanations for lunar craters. However, we now realise that the Moon's large craters are mostly billions of years old. They have remained almost pristine because its airless and waterless surface lacks processes that would wear them away, whereas erosion and other resurfacing processes have removed most traces of the impact craters that scarred the early Earth.

Acceptance of the impact origin of lunar craters received a major boost in 1960 thanks to the work of Gene Shoemaker (1928–97), who studied a 1.1-kilometre crater in Arizona, nowadays usually called 'Meteor Crater'. Shoemaker found minerals that can form only under high pressure, plus other features that demonstrated it had been excavated by shock waves like those from the impact of a body hitting the ground at a speed of tens of kilometres per *second*. This is indeed the orbital speed of objects in the inner Solar System, and such pressures cannot be generated in a volcanic explosion. We are now certain that this particular crater was made by the impact of a meteorite about 50 metres in diameter some 50,000 years ago.

The realisation that any crater resulting from such a 'hypervelocity' impact is excavated by shock waves avoided

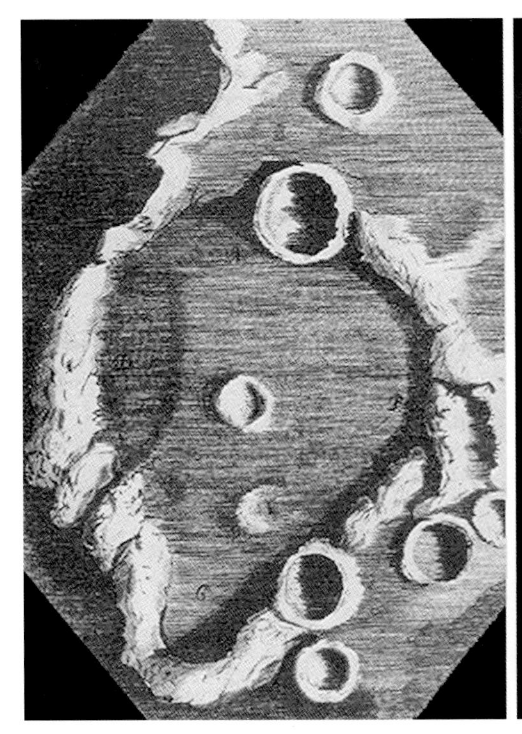

*The first known drawing of a single feature on the Moon, the 150-km diameter crater now named Hipparchus, as published in 1665 by Robert Hooke (*Micrographia *Plate 38)*

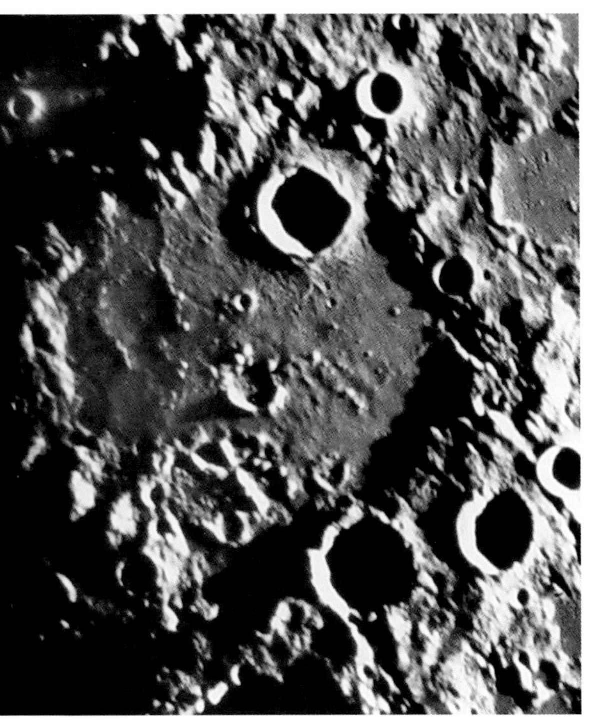

The same area photographed through a modern telescope while the Moon was illuminated almost as it had been at the time of Hooke's drawing Consolidated Lunar Atlas

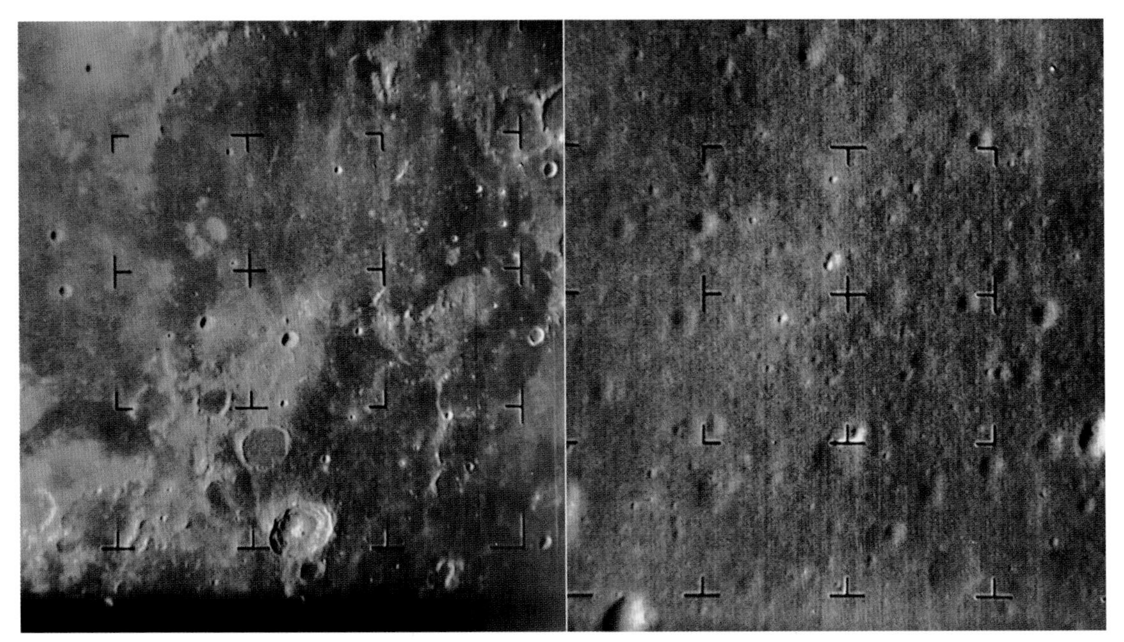

Two views of the Moon transmitted by Ranger 7 in July 1964 on its way to crash into the surface. View from 1,170 km

View from 6.5 km showing an area only 2.7-km wide, in which the smallest visible crater is only a few metres across

one of the principal objections to the impact theory. Almost all of the Moon's craters are roughly circular, with very few elongated examples. Impacts ought to occur randomly at all angles, and it used to be argued that oblique impacts ought to leave a substantial proportion of elongated craters. In fact, the shock waves that do the excavating radiate equally in all directions, irrespective of the direction in which the impactor was travelling. Thanks to further studies, we now know that craters are typically about 30 times wider than the impactor that caused them, and that the impactor is destroyed by the shock waves that excavate the crater – which is why we don't find obvious lumps of the impactor sitting on the crater floor.

When we saw our first close-up views of the Moon from space probes (see previous page), it became obvious that the Moon has craters across a continuous size range down to the limit of visibility. There is no plausible way to make volcanic craters across such a wide size range. This swung the argument firmly in favour of impacts because impactors can be of any size. We also now know from direct observations of asteroids, which form the majority of potential impactors, that their numbers increase as their size decreases in the same manner that the Moon has vastly more smaller craters than larger ones.

Basic lunar geography

Let's now pause to establish some basic information, before describing its relevance for scientists. The near side of the Moon, the side that permanently faces Earth, is notable for having two dominant types of terrain (shown above). The first comprises dark (low albedo) patches known as maria. This is Latin for 'seas', a term bestowed by seventeenth-century astronomers. Most realised these could not be actual water-filled seas because the craters showed they are solid rather than liquid, but took them to be the floors of dried-out oceans. Maria is pronounced 'MAH-ri-a' (not like the female name 'Ma-RI-a') and the singular form, mare, is pronounced as two syllables: 'MAH-ray'. The second type of terrain is the bright (high albedo) ground that surrounds the maria. Being the counterpart to the maria, this was initially dubbed the 'terra' (Latin: land) but that term has fallen out of use and has been largely superseded by 'lunar highlands'.

Until the space age, nobody knew whether or not the Moon's far side resembled the near side. As soon as the Soviet Union's probe *Luna 3* flew past the Moon in October 1959 and transmitted the first grainy pictures of the far side, it became obvious that the two sides were different. As the much clearer view above shows, the far side is predominantly highlands with only a few small patches of

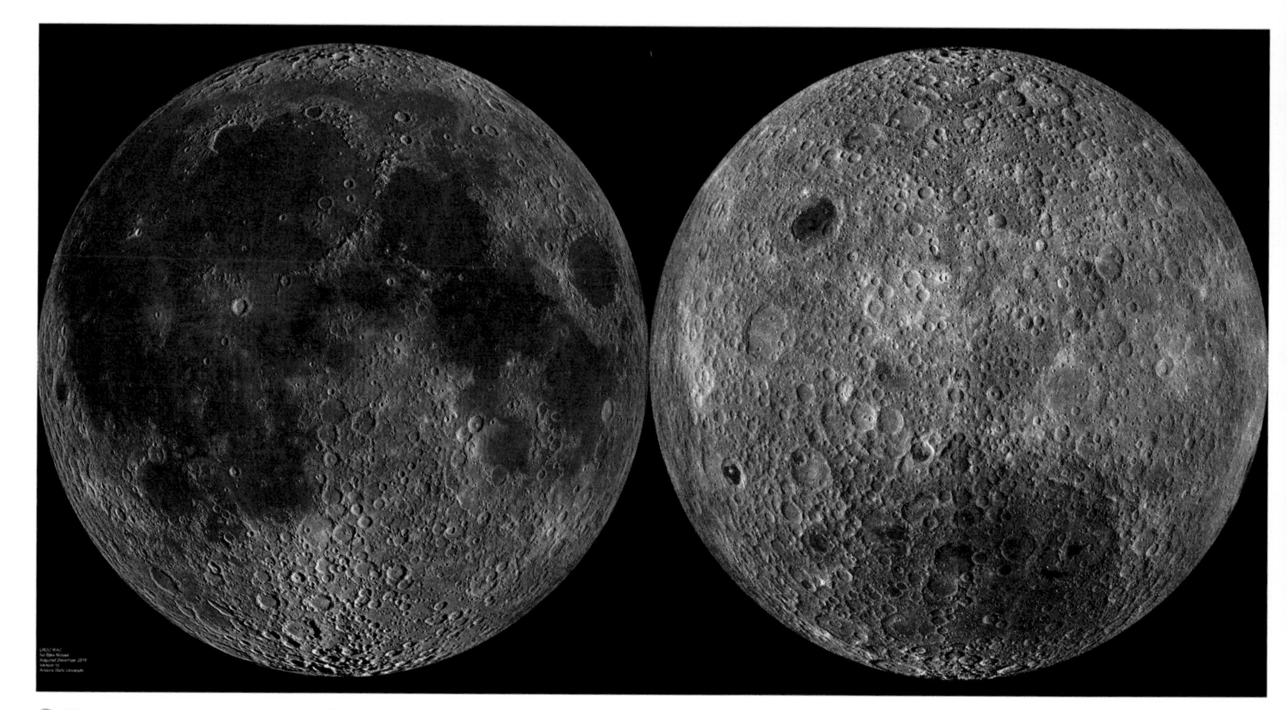

These are mosaics constructed from images obtained by NASA's Lunar Reconnaissance Orbiter Camera (LROC). Left: the near side of the Moon, right: the far side

maria. The far side does have the same units as the near side, but they are present in different proportions. It is apparent on both sides of the Moon that the density of craters on the maria is less than on the highlands. Once you accept that craters are made by impacts, the simple conclusion is that the highland surfaces are older and have been battered by impacts for longer than the younger maria. The maria have covered parts of the highlands, hiding most of the craters that were already there.

We now know that the maria are vast fields of lava that erupted into the largest craters (usually called impact basins). On the near side, this lava completely filled some basins, spilling over their rims and resulting in irregular outlines, whereas less-deeply flooded maria are clear circular patches. On the far side, some basin floors had no eruptions at all and others became only partially flooded. This is mostly because the lunar crust is thinner on the near side, allowing lava to erupt more easily.

Scientific mapping

As Megan Barford discusses in her essay, telescopic observers began drawing general maps of the Moon's near side about the same time as Hooke was making his detailed drawing of a tiny part of it. These were essentially exercises in cartography intended to document the surface features, which for convenience required many of them to be given names. By contrast, geologists make maps to help them understand the history of a region, by identifying tracts of terrain of which the ages (and possibly origin and composition) can be shown to be different. This is usually achieved by means of stratigraphy – working out what is on top of what. The first published attempt is shown on page 184, which is based on a 1960 study by Robert J. Hackman and Arnold C. Mason. It shows the maria (here regarded as all of the same age) in yellow, overlying 'pre-maria' terrain (the lunar highlands) in brown, and what they labelled 'post-maria rocks' (interpreted as craters and their ejecta that were formed after the maria) in green.

With only three time-divisions, Hackman and Mason's map was very basic, and was soon superseded by more detailed studies. In 1961 Gene Shoemaker (of Meteor Crater fame) circulated his prototype geology map of the area around the crater Copernicus shown at the top of page 185. Shoemaker divided the materials he could see on the surface into five age classes and further subdivided some of these according to texture (such as smooth versus hummocky). He offered interpretations for each of his units, such as 'probably chiefly crushed rock' (for crater ejecta) and 'probably volcanic flows' (for mare material). He even added a couple of cross-sections indicating inferred sub-surface structure to a depth of several kilometres.

Shoemaker collaborated with two colleagues to publish a more detailed map of the same area in 1967, shown at the bottom of page 185. This was based on photographic images obtained by ground-based telescopes, supplemented by visual observations made using the 24-inch Lowell refracting telescope at Flagstaff, Arizona, to check fine details. The first-named author of this map was H.H. Schmitt, more commonly known as Harrison 'Jack' Schmitt who, as the Apollo 17 Lunar Module Pilot in 1972, became the first – and to date, only – professionally qualified geologist to walk on the Moon.

This sort of mapping was a necessary precursor to the Apollo programme that successfully landed a total of 12 astronauts in six out of seven attempts. A lot of preparation by way of detailed mapping was required before such a risky venture could be attempted. Much of the effort was directed towards identifying and avoiding landing sites that would be hazardous. It also provided a basis for understanding each lunar region sufficiently well for the experiments performed on the surface and samples brought back to give the best scientific returns.

Lunar missions

The series of geological maps exemplified by the map at the bottom of page 185 was the last to rely only on what could be seen from Earth. American and Russian probes that were sent in the Sixties to crash, soft land or orbit the Moon imaged small areas in detail, while NASA's Lunar Orbiter project placed five uncrewed probes in orbit around the Moon in 1966–67, imaging 99 per cent of it at a resolution of 60 metres or better.

After the Apollo programme wound up in 1972 and the final sample returned from the Soviet crewed Luna programme in 1976, there was a long hiatus before a new generation of uncrewed missions (mostly orbiters) from several nations and agencies. Highlights of the history of lunar missions are listed in the table on page 186.

We can now expect renewed activity by NASA (whose lunar interest is both in scientific missions and as a possible stepping-stone towards human exploration of Mars), collaborative science missions between the Russian space agency (Roscosmos) and the European Space Agency, privately funded rovers, and maybe even a crowd-funded lander (Lunar Mission One) intended to drill the lunar South Pole.

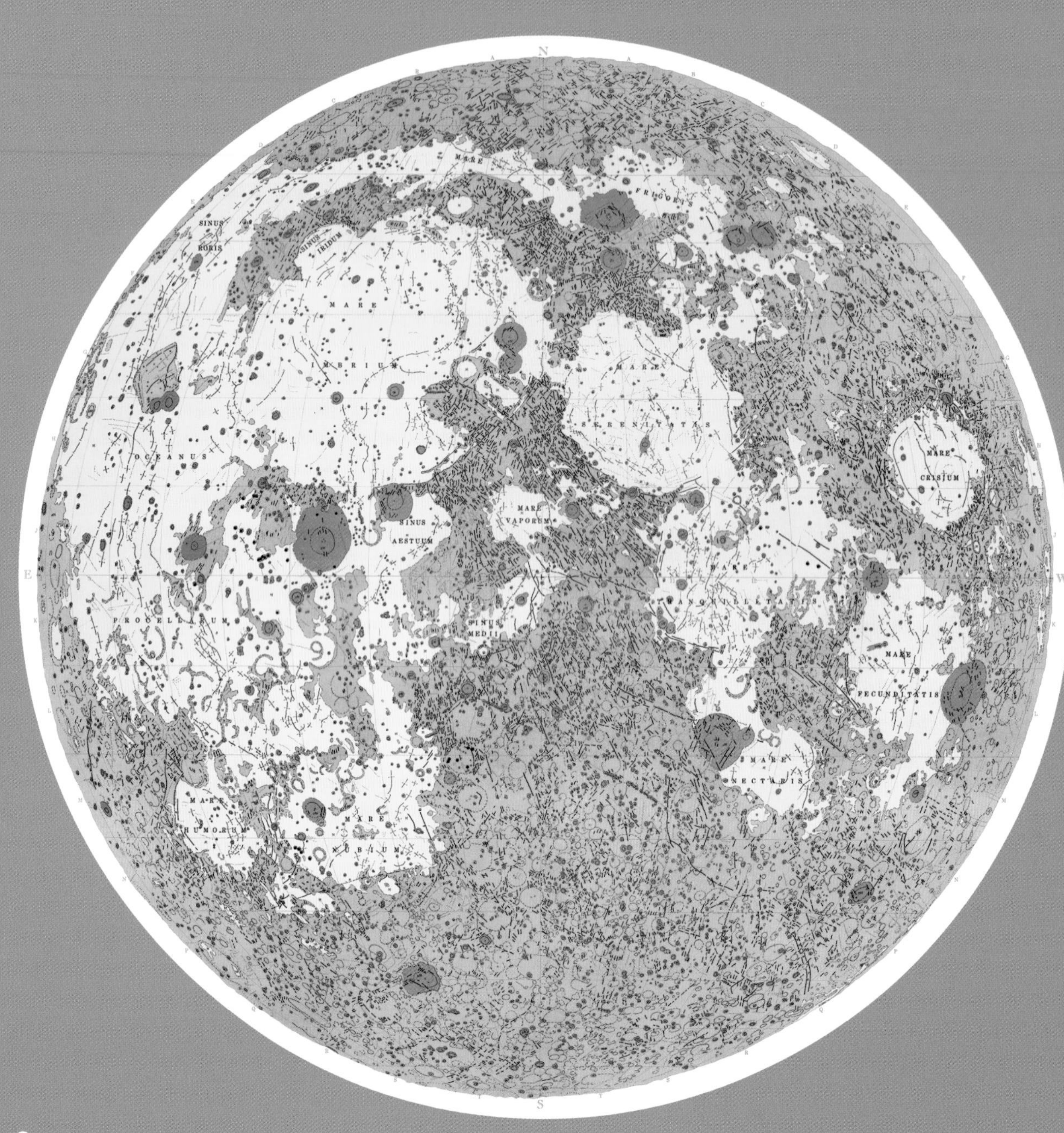

Hackman and Mason's 'Generalized Photogeologic Map of the Moon'

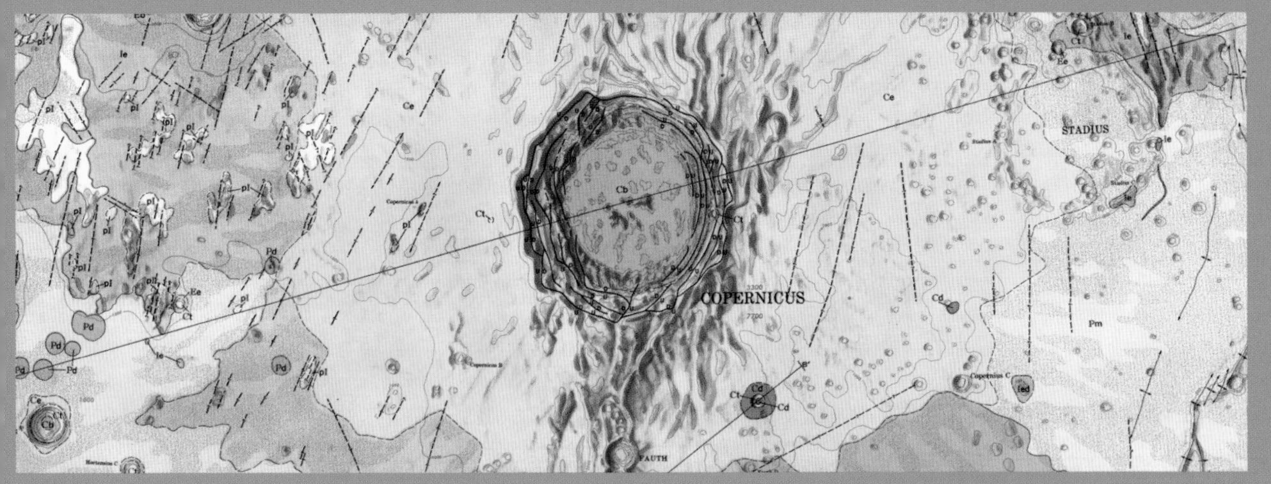

A more detailed view from Shoemaker's 1961 prototype Copernicus geologic map

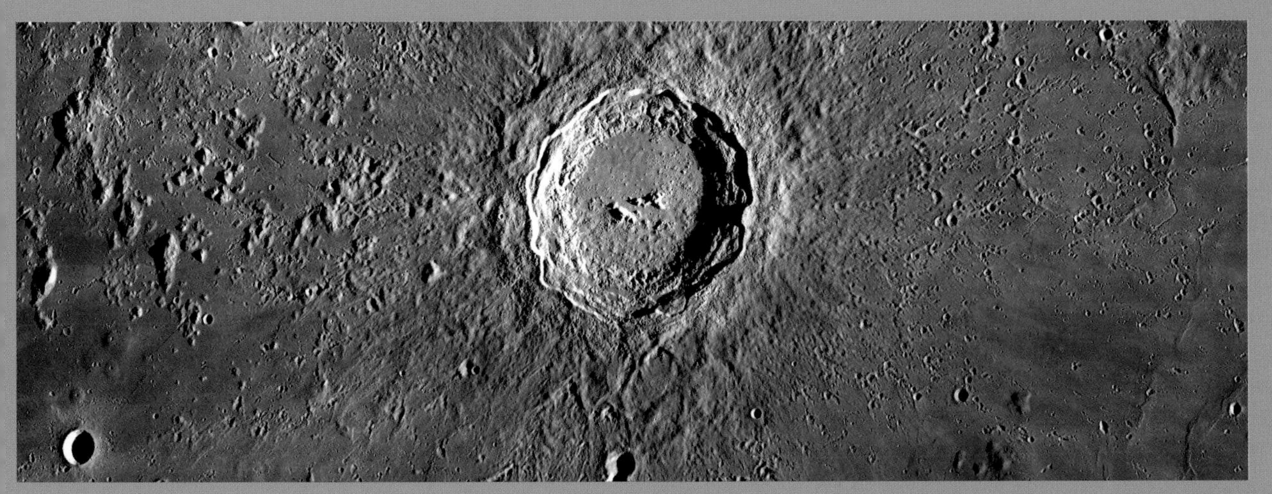

The same detailed area as shown in the maps above and below constructed from images obtained by NASA's Lunar Reconnaissance Orbiter Camera

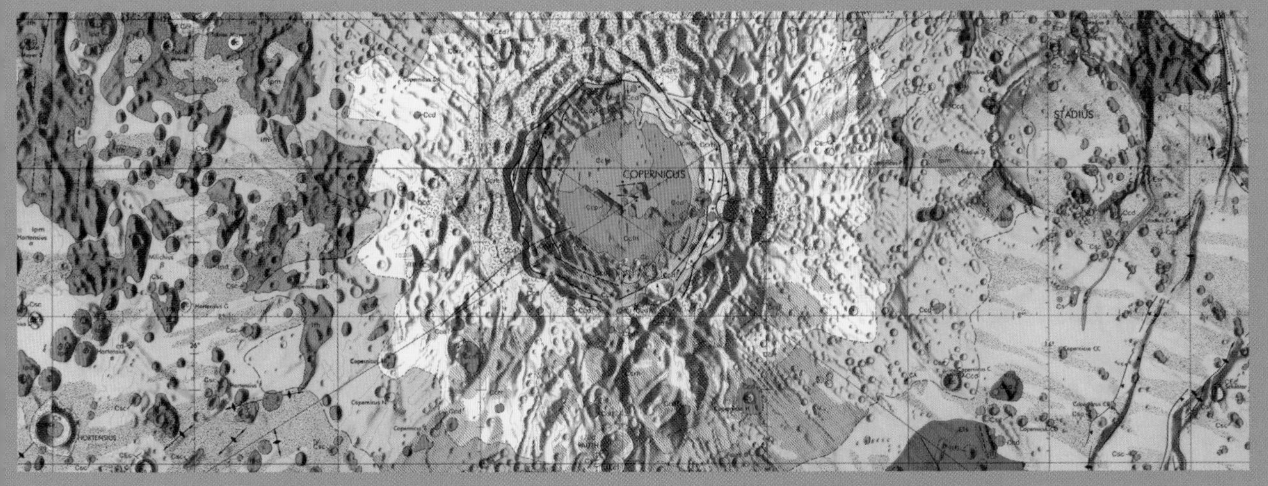

Schmitt, Trask and Shoemaker's 1967 Copernicus geological map showing the same area as the older map above, distinguishing more units and with more detail in its cross-sections

Highlights of lunar exploration

Name	Country	Date	Achievements
Luna 1	USSR	4 Jan 1959	Fly-by, no pictures
Luna 2	USSR	13 Sep 1959	Impact onto Moon
Luna 3	USSR	6 Oct 1959	Fly-by; first far side pictures
Ranger 7	USA	31 Jul 1964	Impactor
Luna 9	USSR	3–6 Feb 1966	Lander; first pictures from the surface
Luna 10	USSR	3 Apr–30 May 1966	First lunar orbiter
Surveyor 1	USA	2 Jun 1966–7 Jan 1967	Lander
Lunar Orbiter 1	USA	14 Jun 1966–29 Oct 1967	Orbiter
Apollo 8	USA	24–27 Dec 1968	First crewed orbiter
Apollo 11	USA	20–21 Jul 1969	First crewed landing; 21.5 kg of samples
Apollo 12,14–17	USA	Nov 1969–Dec 1972	Crewed landings; 360 kg of samples
Luna 16	USSR	20–24 Sep 1970	First robotic sample return (0.1 kg)
Lunokhod 1	USSR	17 Nov 1970–14 Sep 1971	First lunar rover; 11.5 km traverse
Luna 20, 24	USSR	Feb 1972, Aug 1976	Robotic sample returns (0.2 kg)
Lunokhod 2	USSR	15 Jan–11 May 1973	Lunar rover; 40 km traverse
Hiten	Japan	Mar 1990–Apr 1993	Orbiter/impactor
Clementine	USA	Feb–June 1994	Orbiter; mapping including laser altimetry
Lunar Prospector	USA	Jan 1998–Jul 1999	Orbiter/impactor; chemical mapping; found signs of polar ice
SMART-1	Europe	Nov 2004–Sep 2005	Orbiter
SELENE (Kaguya)	Japan	Oct 2007–Jun 2009	Orbiter/impactor
Chang'e 1	China	Nov 2007–Mar 2009	Orbiter/impactor
Chandrayaan-1	India	Nov 2008–Aug 2009	Orbiter and impactor; mineralogical mapping
Lunar Reconnaissance Orbiter	USA	Jun 2009–present day	Orbiter
LCROSS	USA	9 Oct 2009	Impactor; confirmed ice in permanently shadowed crater target
GRAIL	USA	Jan–Dec 2012	Gravity-mapping from twin satellites
LADEE	USA	Sep 2013–Apr 2014	Orbital study of lunar exosphere
Chang'e 3	China	Dec 2013–Jul 2016	Orbiter, lander and rover (Yutu)
Chang'e 4	China	2019	Orbiter, far side lander and rover (Yutu 3)
Chandrayaan-2	India	2019*	Orbiter, lander and rover
Chang'e 5	China	2019*	Sample return mission

** dates subject to change*

Moon rocks

Bringing samples back to Earth, where they can be subjected to detailed and precise analysis, was the most important scientific goal of the Apollo project. These samples, plus observations by the astronauts themselves, confirmed that the maria are covered in volcanic lava made of basalt, whereas the lunar highlands are anorthosite rock (made largely of a variety of feldspar called anorthite). By measuring the accumulated products of radioactive decay within rocks and minerals, it has also been possible to calculate their absolute age and relate this to the relative age scale that had been constructed by counting the density of craters in different regions. This shows that the rate of crater formation was intense during an event 4.1–3.8 billion years ago known as the late heavy bombardment. That phase of cratering obliterated older craters, so we don't have direct evidence of whatever happened before then. The term 'late' denotes late relative to the formation of the Solar System, although it was early in lunar history (the Moon and Earth being about 4.5 billion years old). Since about 3.5 billion years ago, the rate of cratering seems to have been fairly constant and far less intense than during the late heavy bombardment, though it is impossible to rule out short flurries of bombardment.

Thirty known basins were formed during the late heavy bombardment. The South Pole Aitken basin in the south of the far side is the oldest, perhaps formed about 4.1–4.0 billion years ago. It occupies the central part of the southern hemisphere of the far side as shown in the image at the bottom of page 182, but is not easy to see because of the lack of dark mare-fill. The Imbrium basin in the north centre of the near side was formed about 3.8 billion years ago, and the Orientale basin (the youngest of its kind) is about 3.7 billion years old.

Surprisingly, in all cases the mare basalts are several hundred million years younger than the basins where they occur. This age difference (more than twice the interval between the dinosaurs and us!) shows that the mare basalts cannot simply be made of magma that escaped from eruptions triggered by basin-forming impacts, or be surface rock that was melted by the heat of the impact. Although most mare basalts had been emplaced by about 3 billion years ago, smaller patches of lava continued to erupt in some areas until about 1 billion years ago.

Other analyses of lunar samples include determining what minerals make up the rock and whether the abundances of chemical elements differ from those in rocks from Earth. For elements such as oxygen that have more than one stable isotope, the relative abundances of these isotopes can be used as a 'fingerprint' to indicate whether the Earth and Moon were made from the same source material. You can learn things about a rock's origin and history that cannot be deduced from orbit even by very basic visual examination of a lump of it, or studying a thin slice of it under a microscope (shown overleaf).

The six Apollo landings collected 382 kilogrammes of samples between them, while the three Soviet Luna sampling missions brought back a further 0.3 kilograms. All of these were from the near side, so some important regions remain unsampled. The South Pole Aitken basin is perhaps the most compelling target for future sampling, because we may find material from deepest inside the Moon there.

An unexpected and very important boost to our Moon-rock collection began in 1982 when a meteorite collected from the ice in Antarctica attracted attention because it was unlike the normal kinds of meteorites, which are fragments from asteroids. If there hadn't been samples from the Moon with which to compare it, its origin might still be in doubt, but it was soon realised that this one closely matches some of the Apollo samples and must have been thrown into space as a fragment of ejecta from a lunar impact crater. Several other lunar meteorites were then tracked down in existing collections and many more were subsequently collected by searching in Antarctica and other barren places. Over 200 kilogrammes of lunar rock has now been collected as meteorites. We cannot identify the specific location from which any lunar meteorite originated, but common sense tells us that about half of what we have is likely to have come from the far side.

Some lunar rocks are breccias, made of fragments of basalt and/or highland rock that were smashed by impacts and then welded together by the heat generated. Other samples are lumps of a single basalt (such as those shown opposite) or are highland rock type. The lunar soil or 'regolith' contains fragmented or pulverised relics of all of these, as well as glassy beads about 0.1 millimetres in size that appear to be frozen droplets of basaltic spray from explosive eruptions.

Most mare basalts in and around Mare Imbrium show geochemical fingerprints of having been extracted from source rocks that were relatively rich in potassium (K), rare earth elements (REEs) and phosphorus (P), and so are

A specimen of mare basalt collected by Apollo 15 astronauts, on a sample table. The cube next to it is 2.5 cm wide. The holes were made by gas escaping from the lava when it was still molten

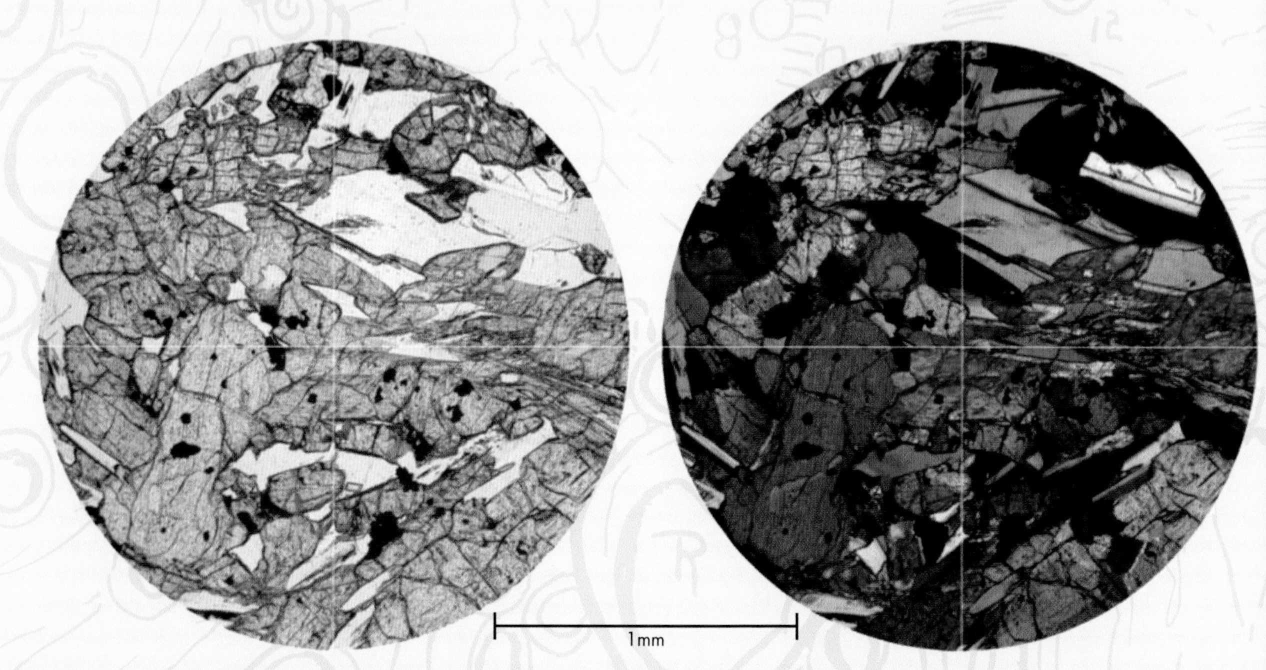

Two views of the same area of a thin slice of the sample shown above (avoiding the bubbles) seen through a microscope: in plain light (left) and using polarized light to reveal mineral properties by means of interference colours (right). In the left view, the black areas are iron-titanium oxide (ilmenite), feldspar is clear, and olivine and pyroxene are grey. In the right view, olivine and pyroxene show strong colours, whereas the feldspar is grey and stripy

known as KREEP. It is thought that this area has sampled an anomalous patch in the nearside interior where excess heat from radioactive decay of potassium and thorium (mapped from orbit by detecting their emitted gamma rays) was the main cause of the melting that led to the eruption of the mare basalts. So, as well as the relative thinness of the near side crust we can add locally stronger radioactive heating as a factor contributing to the hemispheric asymmetry in the distribution of the maria.

Below the Moon's surface

The Moon is less dense than Earth, which shows that its interior must be mostly rocky without a large iron core like Earth's. Apollo 12, 14, 15 and 16 left instruments called seismometers embedded in the lunar surface to record the tiny vibrations caused by moonquakes and transmit the data to Earth. Data from these were collected until September 1977, when reception was discontinued as a cost-saving measure. The Moon is geologically less active than Earth, but hundreds of tiny moonquakes were detected, enabling the Moon's internal structure to be deduced. The surface rock, or crust, is about 50 kilometres thick. A solid rocky mantle makes up the bulk of the interior, and this is the region from which the magma that was erupted as mare basalts was derived. Analysis of the data in the post-Apollo era was unable to determine whether or not there was a central iron core, but reanalysis with twenty-first century data processing techniques suggests a core of 330-kilometres radius and hints that the lowest 150 kilometres of the mantle might be partially molten. This modern concept of the Moon's interior is shown to the right.

Another way to probe the Moon's interior is to map how the strength of its gravity varies from place to place. This was first done by radio tracking of satellites in lunar orbit (including some Apollo capsules) and was able to show stronger than expected gravity over the lunar maria. It was originally suggested that these 'mass concentrations' or 'mascons' were evidence of buried basin-forming impactors, but they were soon interpreted as showing that the crust has been thinned where a basin has been formed, bringing the denser mantle closer to the surface. The eruption of mare basalts into these basins added further dense rock that boosted the local gravity even more.

NASA's GRAIL mission in 2012 deployed twin satellites in tandem in a low lunar orbit about 55 kilometres above the surface. These communicated to each other using microwave signals to measure their separation to a precision of better than one micrometre, enabling a much more detailed gravity map to be made (see page 190). As well as confirming the mascons associated with most maria, it also revealed some narrow linear gravity anomalies that might represent the fractures through which much of the mare basalt erupted.

Wet or dry Moon?

The analytical techniques available in the Apollo era failed to detect water bound up inside lunar rocks or chemical alteration of minerals caused by water. Lunar mineral crystals that are billions of years old look as fresh as if they grew only last year; on Earth, the damp environment would have led to chemical alteration penetrating along fractures and cleavage planes.

The Moon is essentially a dry place, but it has turned out not to be bone dry after all. This century, sensitive analytical techniques have found traces of water in Apollo samples within grains of a mineral called apatite. Apatite's crystalline structure causes it to scavenge water, and also any large negatively charged ions such as those formed by fluorine and chlorine, from the magma within which

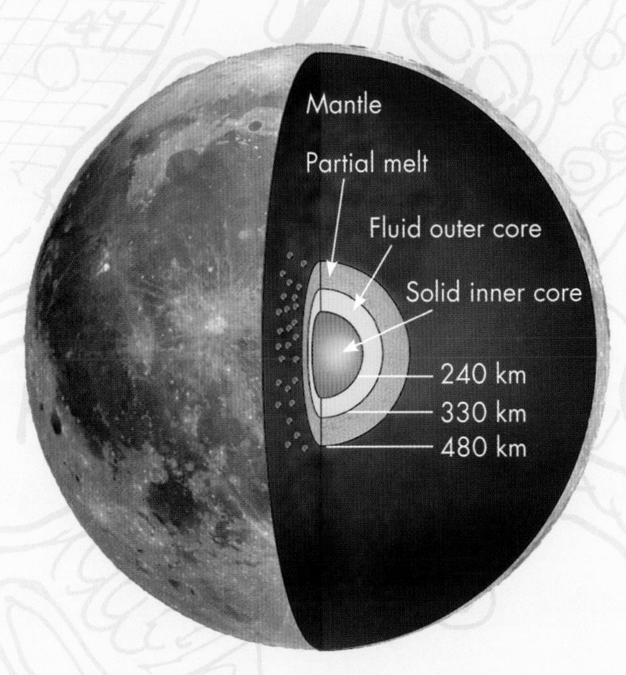

The Moon's interior based on a modern reanalysis of data collected by the Apollo seismometers. The dots show where most of the detected moonquakes originated.

Two gravity maps of the Moon's near side produced from data collected by the twin GRAIL orbiters and superimposed on the near side image from page 182. Left: gravity anomalies (red = stronger than expected, blue = weaker than expected).

it grows. There is an ongoing debate about how to use apatites to interpret the water content of the Moon's mantle, but it could be anything from ten parts per million to one part per thousand. We aren't sure what the average water content of the Earth's mantle is either, but it is almost certainly higher than the Moon's, perhaps about 1 per cent.

The Moon's surface rock and its interior are still regarded as much drier than Earth. However, there is also an unrelated reservoir of water on the Moon. This occurs due to ice inside the cold (-170 °C) craters near the poles, whose floors are never illuminated by the Sun. Evidence for this built up slowly. In 1994, the *Clementine* orbiter showed that radio waves bounce from those craters in a way consistent with the presence of ice. Five years later, the neutron spectrometer on the *Lunar Prospector* orbiter demonstrated concentrations of hydrogen, which could most reasonably be interpreted as residing in water (H_2O). The clincher came in 2009, when the *Centaur* rocket that had delivered the *Lunar Reconnaisance Orbiter* was crashed into the permanently shadowed floor of a crater named Cabeus near the Moon's South Pole. The *LCROSS* (Lunar Crater Observation and Sensing Satellite) probe trailed six minutes behind, and before it too crashed, it had time to analyse the ejecta flung up by the previous crash,

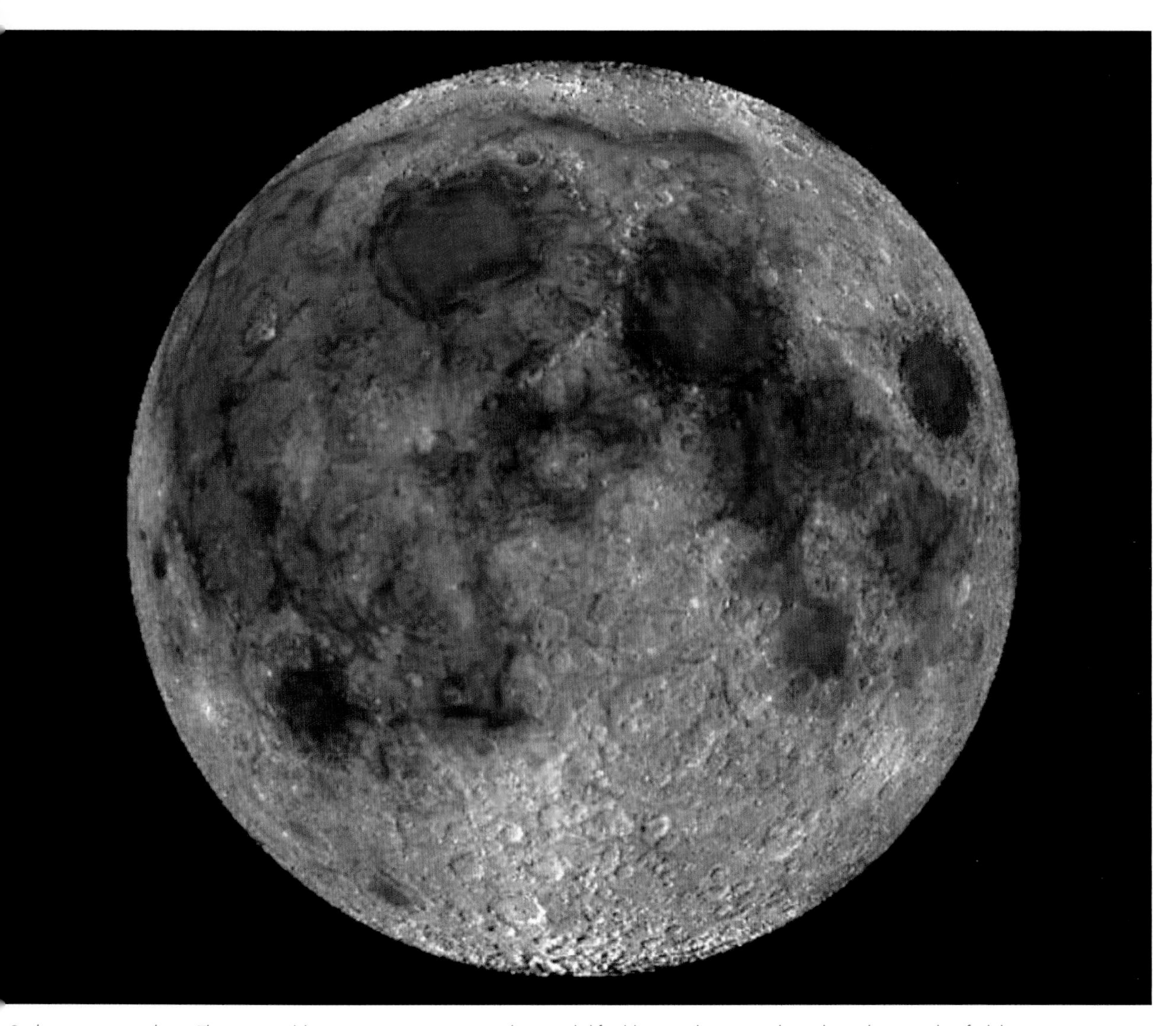

Right: gravity gradient. The narrow blue zones may represent dense solidified lava within vertical cracks in the crust that fed the eruptions that produced the maria.

showing the floor of the Cabeus to consist of about 6 per cent water by mass.

There is a fairly simple explanation for the Moon's polar ice, and it is nothing to do with the water from inside the Moon that has been there since the Moon's birth. When a comet hits the Moon and makes a crater, the comet's ice is vaporised, liberating water molecules. If a molecule hits a hot surface, it will bounce and eventually be lost to space. However, if it hits a cold surface, it will stick there. If it hits a permanently cold place, it will stick there indefinitely. In this way, permanently shadowed crater floors act as 'cold traps' where ice accumulates molecule by molecule.

The Moon's origin

Scarcity of water is only one example of how the Moon differs from the Earth. It has generally lower abundances of 'volatile elements' such as sodium. On the other hand, the relative abundances of the three stable isotopes of oxygen in lunar rocks match almost perfectly what we find in the Earth, which suggests that the two bodies formed from the same source material. However, if that were the case, why does the Moon have a comparatively smaller iron core than the Earth? These are apparent contradictions that have to be reconciled when trying to explain how the Moon formed.

In 1879 George Darwin (1845–1912), second son of his more famous father Charles, proposed that the Moon formed by splitting off from the outer part of the Earth. Darwin lacked the data to test his idea, but we now see that it could explain the Moon's small core and its oxygen-isotope match to Earth's, though not its depletion in volatiles.

A model in which Earth and the Moon grew side by side while the Solar System was forming falls down in relation to the core because each should have the same proportion of core-forming material. Capture of the Moon by Earth after independent formation would be dynamically very difficult and would require the matching oxygen isotopes to be a fluke.

The Moon's origin is still the subject of scientific debate, but a theory that it was formed by a 'giant impact' onto the early Earth is now widely accepted. This was developed in the mid-Eighties because it could explain the matches and mismatches between the two bodies that Apollo had revealed. The late stages in the growth of terrestrial planets are probably a series of a few collisions (giant impacts) between bodies of roughly similar size (known as planetary embryos), rather than growth of a larger body by mopping up numerous much smaller bodies. Giant impacts usually result in two planetary embryos merging, leaving a larger embryo that would be predominantly molten because of the heat generated by the impact. This melting would make it easy for iron (which is dense) to sink inwards to form a core.

According to the giant impact theory for the Moon's origin, the final giant impact experienced by the proto-Earth was when a Mars-sized planetary embryo struck it a glancing blow. Rather than leading to full merger, the collision ejected part of the impactor's mantle and some of the Earth's mantle into space as hot debris from which most of the volatiles escaped. Meanwhile, the impactor's core ploughed inwards and merged into the Earth's core. The Moon then grew in orbit round the Earth from the volatile-depleted debris.

Energy converted to heat as the Moon grew would have melted it, allowing what little free iron there was to sink and form a tiny core. More importantly, as the Moon's global magma ocean began to cool, the first crystals to form would have been anorthite, the very mineral known to make up most of the lunar highlands. Anorthite has a relatively low density, so these crystals would tend to rise. They would probably need to clump together to form larger masses before their buoyancy overcame the magma's viscosity, but then these 'rock-bergs' would rise until they bobbed up to the surface where they coalesced to form the lunar highland crust see opposite page.

Radioactive dating has determined the oldest lunar crust samples to be about 4.35 billion years old. This is about 200 million years after the birth of the Solar System, but the Moon-forming impact could have occurred as long ago as about 4.5 billion years. Water and individual volatile elements would have been preferentially lost to space before the debris from the giant impact was able to coalesce into the Moon, so that the giant impact hypothesis is the most credible story. The giant impactor has been given the name Theia, after the mother of Selene, the Moon goddess of Greek mythology. A recent variant of the hypothesis, invoked to explain some of the differences between the near and far sides is that the giant impact debris originally formed two moons, which merged in a relatively slow mutual collision a few tens of millions of years later.

What next for the scientists and the Moon?

Whether any of the current variants of the giant impact hypothesis for the Moon's origin survives the test of time remains to be seen. There is much still to be learned in other aspects of lunar science too. Most of the Apollo lunar samples remain untouched, having been kept pristine for future analysis as techniques and ideas advance. However, it is undeniable that scientists are hungry for more, especially from hitherto unsampled regions such as the far side, the poles and the youngest mare lavas on the near side. We don't really understand the causes of the internal melting episodes that led to mare basalt eruptions hundreds of millions of years after the formation of the basins they now occupy. Scientists would also like to deploy more instruments on the lunar surface; for example, to measure the rate of heat flow from the interior and to detect and study moonquakes.

There is probably a record of how the solar wind has changed over time trapped within regolith buried under lava flows, which could help us understand the Sun's behaviour much better. There may even be pristine fragments from the Earth's early crust (which no longer survive at home) that fell to the Moon after being blasted off the Earth by major impacts. Engineers too are keen to learn more about the Moon to identify resources that could be used to build or supply lunar bases without having to ship everything from Earth.

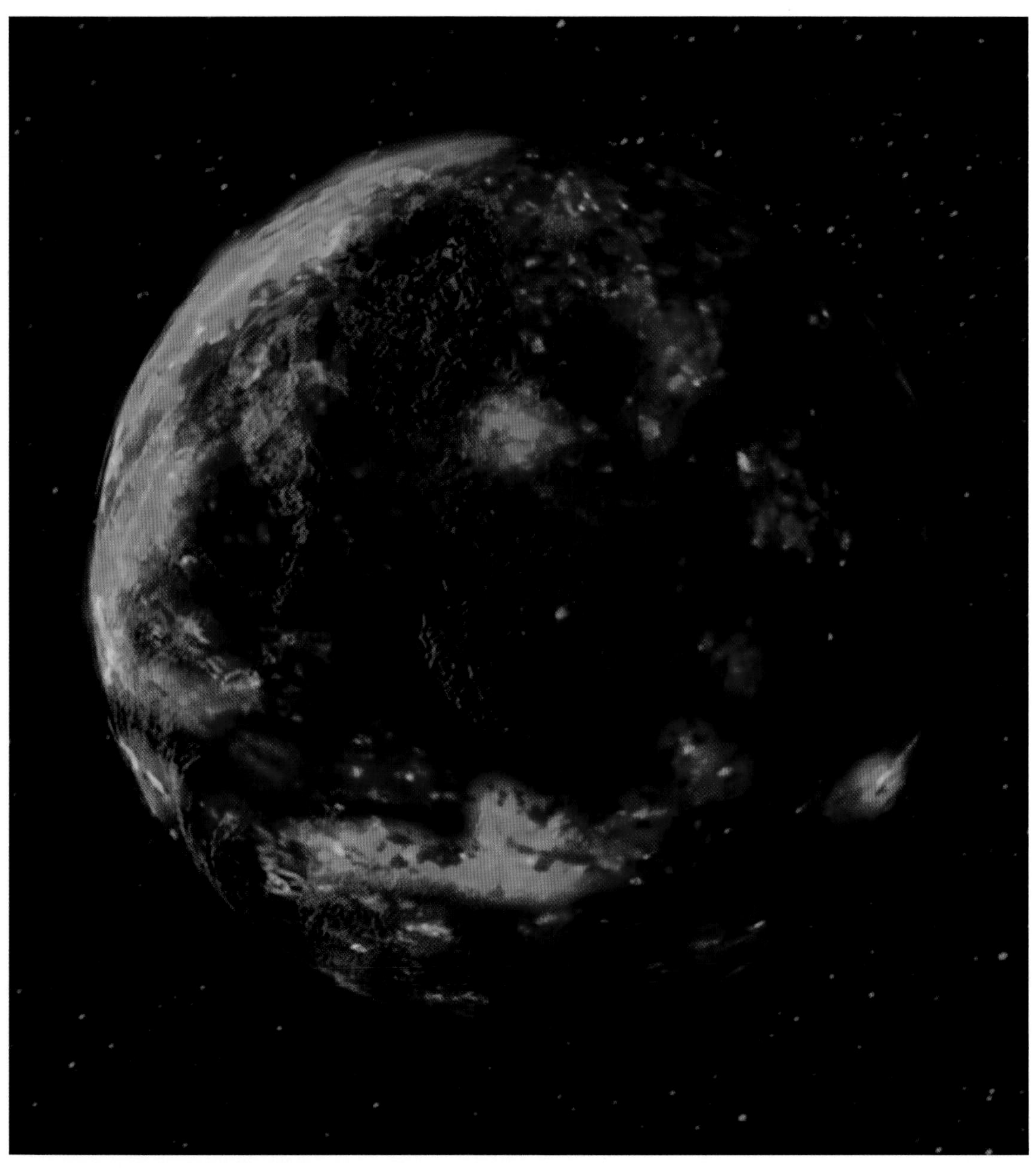

Artist's impression of how the young Moon may have looked, with large bodies of anorthite having risen to the surface of a magma ocean

So What If It's Just Green Cheese? The Moon on Screen

Simon Guerrier

In the years leading up to the first Moon landing, science fiction seemed to eagerly anticipate our escape from the confines of Earth. Stanley Kubrick's *2001: A Space Odyssey* (1968) showed spaceflight in painstaking detail and with awesome effect. Yet when Dr Heywood Floyd takes a commercial flight up to Space Station V on his way to Clavius Base on the Moon, the journey has become so routine that he barely glances out the window at the spectacular view. In two *Doctor Who* stories – 'The Moonbase' (1967) and 'The Seeds of Death' (1969) – the time-travelling Doctor visits future lunar colonies whose personnel control Earth's weather. The former seems optimistic; the all-male staff from different nations united in common cause. But in the latter, broadcast just months before the real Moon landing, a lone enthusiast complains that, 'Nobody cares any more about exploring space'. We learn that humanity never got beyond the Moon and 'it's been years since we sent up a satellite'. In the weeks that this story was broadcast in the UK, US network NBC cancelled *Star Trek* (1966–69). Even though that series rarely mentioned the Moon, it presented a utopian vision of men and women from all nations and races working together to explore deep space – the great leap that would follow the impending first small step. However, despite a dedicated following (among 'the upper middle class and highbrow audiences', as one newspaper put it), *Star Trek* had failed to find favour with a wider audience.[1] Even so, when it debuted in the UK on 12 July 1969, it was part of the BBC's package of programmes to engage the general public with the real-life Apollo 11, which launched four days later.

Both the BBC and ITV tried to attract a wide audience for the historic event. On 20 July, the day of the landing, BBC One dovetailed coverage around perennial Sunday-night favourites. *Songs of Praise* (from the US Airbase in West Ruislip) was followed by a live update from Apollo 11, as astronauts Neil Armstrong and Buzz Aldrin began their descent. The usual Sunday night schedule then continued – an episode of medical drama *Dr Finlay's Casebook* (in which a male doctor tried to get rid of a female colleague) and *The Black and White Minstrel Show*. Then it was back to Apollo 11, with the astronauts' voices transmitted live as their lunar module landed in the Sea of Tranquillity at 9.18 p.m., UK time. James Burke and Patrick Moore in the BBC studio in London and Michael Charlton reporting live from Mission Control in Houston provided technical explanation.

After a news bulletin, there was a change of tone. In *So What If It's Just Green Cheese?*, Moon-related poetry and excerpts were read by distinguished actors such as Ian McKellen and Judi Dench, with music ranging from a live jam by psychedelic rock band Pink Floyd, Tamás Vásáry playing Debussy's *Clair de Lune* piano sonata and a comic performance from the Dudley Moore Trio. At some point in the evening, a clip from a new single released just nine days earlier was first heard on TV – 'Space Oddity' by the then little-known David Bowie. Viewers were returned to Burke and Moore to watch Armstrong's first steps on the Moon, with coverage continuing into the early hours.

Meanwhile, Dudley Moore's comic partner Peter Cook was live on ITV as part of *David Frost's Moon Party*, alongside guests including actress Sybil Thorndyke, singer Cilla Black and comedian Hattie Jacques. The eclectic mix of music and comedy ran for 10 hours until 3 a.m. Science-fiction author Ray Bradbury found the whole thing too frivolous and walked out of the studio, but the programme attempted some serious discussion. Around midnight, the guests debated the value of lunar exploration. According to Michael Billington in *The Times*, eminent historian A.J.P. Taylor and actor and singer Sammy Davis Junior formed 'a somewhat bizarre alliance in attacking manned space flights' – though Billington concluded that the issue was so fundamental, 'one wished it could have been thrashed out sooner'.[2] We rely on Billington's assessment because we can't watch these programmes ourselves. It's perhaps telling that neither broadcaster felt their coverage of this key moment in human history worth keeping for posterity. (Some home-recorded audio and very brief footage from the BBC coverage survive.)

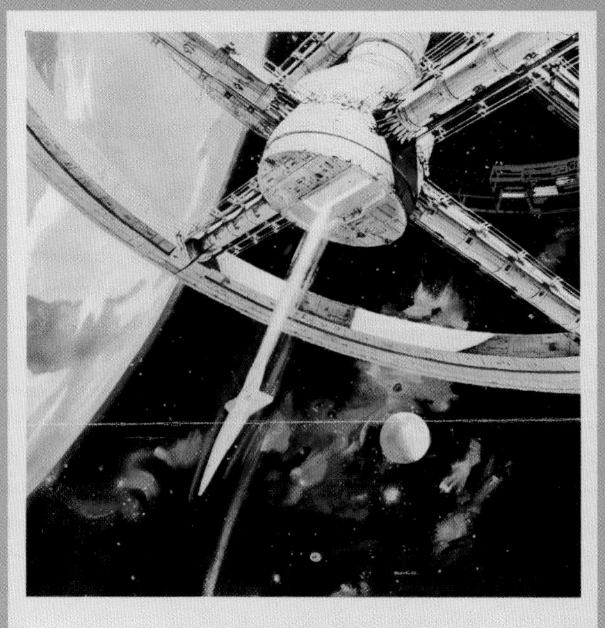

MGM PRESENTS A STANLEY KUBRICK PRODUCTION

STARRING
KEIR DULLEA · GARY LOCKWOOD · SCREENPLAY BY STANLEY KUBRICK AND ARTHUR C. CLARKE

PRODUCED AND DIRECTED BY
STANLEY KUBRICK

SUPER PANAVISION AND METROCOLOR

MGM

Movie poster, 2001: A Space Odyssey *(1968)*

☾ *Sean Connery in* Diamonds are Forever *(1971)*

☾ *The Moon Buggy in* Diamonds are Forever *(1971)*

One programme aimed at a higher-brow audience still exists. On the evening of 21 July, an hour after Armstrong and Aldrin left the lunar surface, the BBC's current affairs programme *Panorama* explored 'The Impact on Earth' in a special live episode. Presenter Robin Day began by saying that 'the most obvious answer' to what had been achieved was prestige for the United States. He then referred to protests held outside Mission Control the previous day to 'remind everyone of the unsolved problems here on Earth'. This led to a report by Julian Pettifer, citing those who saw the landing as 'vanity' and 'folly'. He went on to list the benefits of the space programme: astronaut and satellite pictures of Earth would help find new mineral deposits and oil fields, and help manage agriculture, weather forecasts and road planning. The overall sense is of broadcasters struggling to present the Moon landing as more than a brilliant technical achievement. It was something that had to be made fun and relevant to viewers – even the relatively highbrow audience of *Panorama*.

There's something of this in the fiction that followed, too. *Doppelganger*, released in the UK in August 1969, made the exploration of space relevant by ignoring the Moon and instead sending astronauts to a mirror-image of Earth on the far side of the Sun. *Moon Two Zero*, released in the UK in October 1969 and the US the following spring, was the only science-fiction film to come from Hammer Films – better known for its horror productions. Shot before the Apollo 11 landing, by the time of the film's release its producers seem to have doubted the Moon's allure, advertising the film not in relation to real events but as 'the first western in space'. The ostensibly straight drama is also undermined by the opening title sequence, a daft cartoon in which an American astronaut and Russian cosmonaut battle one another on the lunar surface.

The Clangers debuted on BBC One on 16 November. The animated, mouse-like creatures are visited in more than one episode by astronauts from Earth, but co-creator Oliver Postgate was wary of linking the series too closely to the space programme. He later wrote that had an Apollo mission failed and the astronauts perished – 'as well they might have done' – the series 'would have been one of the blackest jokes in television history'.[3]

Marooned was released in the US in November 1969 and the UK early the next year. The story doesn't involve the Moon, but the plot about three astronauts being stranded in space anticipated events on board Apollo 13 in April 1970. That was also true of 'The Ambassadors of Death', a *Doctor Who* story commissioned before the first Moon landing but broadcast as the crisis unfolded on Apollo 13. The story, set in the near future, is about a British-crewed mission to Mars – not the first one, either – but it owes a lot to the way Apollo was presented on TV. Actor Michael Wisher plays a reporter at Mission Control, softly explaining developments direct to camera, as had the real Michael Charlton in his reports from Houston. Model shots of spacecraft in the eerie blackness of space are accompanied by strange, up-tempo music rather in the style of 'Moonhead', which Pink Floyd performed live for *So What If It's Just Green Cheese?*

In September 1970, ITV began showing adventure series *UFO*, which included a moon base staffed by glamorous, purple-haired women who try to monitor alien invaders. This base was to be established as soon as 1980 but kept secret from the people of Earth. It is perhaps the first touch of paranoia about our presence on the Moon – that the general public were not getting the full story about what had really taken place. It would not be the last. By this time, it was known that the real Apollo programme would conclude with Apollo 17 at the end of 1972. The James Bond film *Diamonds are Forever* (1971) implies the landings were being faked anyway. There's the incongruous sight of Bond (Sean Connery) dressed in an ordinary suit chasing over the lunar surface – really a TV set – and then making off in a Moon buggy. *Silent Running* (1972) is not about the Moon, but its plot about a battle on the spacecraft Valley Forge to protect Earth's last-surviving vegetation is clearly informed by the environmental movement. It's been argued that that movement was galvanised by colour pictures taken of Earth by the Apollo 8 crew in December 1968.

Doctor Who returned to the Moon in 'Frontier in Space' (1973), but no longer took cues from Apollo. The spacesuits in the story owe more to the comic-strip *Dan Dare* and the Moon of the twenty-sixth century is used to incarcerate political prisoners rather than for study. Later that year, the producer and script editor of *Doctor Who* presented *Moonbase 3*, set on one of several lunar settlements in 2003. It aimed to be realistic, and James Burke, who had presented the BBC's coverage of the Apollo missions, acted as adviser. Yet script editor Terrance Dicks later admitted that the time had already passed for such a serious take on the subject and the series failed to find an audience. It also failed to answer the question hanging over coverage of Apollo: why we should go to the Moon at all? In the episode 'Achilles Heel', the director-general back on Earth has to be told what projects the moon base is engaged in – he doesn't already know – and they all sound rather vague. 'So,' he responds, 'the only money-making project you've got is asking for more money'. It hardly seems viable.

An idea for a second series of *UFO* that would focus on its moon base led to the creation of an original series, *Space: 1999* (1975–77). In the opening episode, a nuclear explosion flings the Moon out into deep space – along with the colonists on its surface. Despite the setting, it owes little to Apollo. There's more paranoia in *Le Orme* (1975), an Italian film released in the UK as *Footprints on the Moon*, which suggests strange, secret experiments have been conducted on lunar astronauts. In *Alternative 3* (1977) the Moon and Mars are being covertly prepared for mass-colonisation because climate change is about to make Earth uninhabitable – all presented as a real investigation. *Capricorn One*, released in Japan in late 1977 and in the US the following year, sees the crew of a mission to Mars forced to abandon their mission because of a technical fault. They must then fake a successful mission in a TV studio, as a conspiracy builds. The film owes as much to Watergate as Apollo but surely lends weight to the idea that the Moon landings were also fabricated.

By this time, the success of *Star Wars* (1977) meant film and TV science fiction would largely ignore the parochial details of real space travel in the years that followed. An exception is *Moonraker* (1979), when James Bond himself blasts into orbit. There's a good attempt to make the space travel seem credible. But perhaps it also marks the end of Apollo's influence on how we viewed the Moon. Despite the title, the film takes us nowhere near the lunar surface. Instead, 'Moonraker' is the name of a kind of space shuttle – the craft that succeeded Apollo in fact and aimed much closer to home.

Apollo 13 (1995) dramatises the real events of 1970, while *Apollo 18* (2011) tells a new story rooted in a past, fictional mission that was then kept secret. *Iron Sky* (2012) takes lunar paranoia to the extreme, with the Moon home to Nazis who fled from the Second World War. The acclaimed *Moon* (2009) is yet another conspiracy, anticipating a lunar surface exploited by faceless corporations. It taps into fears about the future of space exploration in private hands – all a long way from Apollo. The cutting-edge technology that first landed people on the Moon has been relegated to history.

Yet the Apollo missions influenced film in less obvious ways, too. In 1976, the Academy Award for Best Cinematography went to *Barry Lyndon*, directed by Stanley Kubrick – who had overseen *2001*. There was nothing sci-fi about *Barry Lyndon*, which was set in the eighteenth century with interior scenes authentically lit by candle. What made that technically possible was that Kubrick and his team used lenses with particularly wide apertures. The Carl Zeiss Planar 50 mm f/0.7 had been designed for the Apollo programme in the Sixties to enable photography of the far side of the Moon.

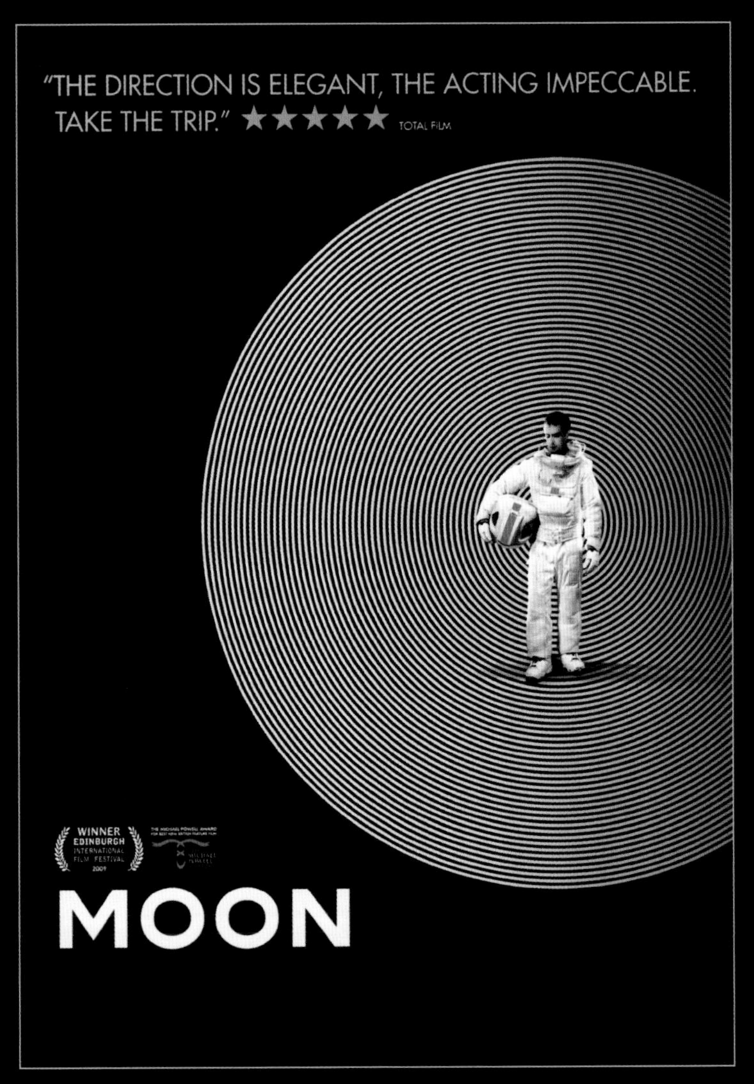

Promotional image from Duncan Jones's Moon (2009)

Poetry and the Moon: Moon Vehicles, Moon Metaphors

Simon Barraclough

If the Moon didn't exist, poets would have to invent it, and in many ways that's what they do with each new poem. There are as many different versions of the Moon as there are poems about it and, though we tend to think we know it very well, the Moon remains our constant mysterious companion.

Poets are intensely interested in the Moon but not particularly interested in walking upon it. They prefer to set their *metrical* feet around it: to approach it and then withdraw; to maintain the provocative distance, the tantalising gap in which desire circulates. This gap is essential to metaphor. Metaphor depends on distance, on non-identity, even while it flirts with collapsing two separate entities into one. Take the title of David Niven's *The Moon's a Balloon* (1971): if Moon and balloon were identical, the metaphor would disappear; it would serve no purpose. For the metaphor to work, the Moon needs to both be a balloon and not be a balloon. To be and not to be, that is the answer.

The first Moon poem that leaps to mind is Sir Philip Sidney's 31st sonnet from his often experimental sequence about unrequited love, *Astrophel and Stella:*

> *With how sad steps, O moon, thou climb'st the skies;*
> *How silently, and with how wan a face.*
> *[...]*
> *O moon, tell me,*
> *Is constant love deemed there but want of wit?*
> *Are beauties there as proud as here they be?*
> *Do they above love to be loved, and yet*
> *Those lovers scorn whom that love doth possess?*
> *Do they call virtue there ungratefulness?* [1]

Look at this constellation of 'O's: the beautiful 'Other' that is the object of desire. After 1969 you might also think, 'One sad step for Moon, one giant leap for Moonkind'. From what I've read about Neil Armstrong, he doesn't seem the type to harbour poetic thoughts and yearnings like Sidney, although they were both explorers, conquerors, servants of ambitious states with big power-moves to make for a watching world.

But Sidney reveals no desire to conquer the Moon, to plant his feet upon its ground. Fondly, he asks this most seemingly inconstant of celestial bodies for reassurance about Earthly love and courtly manners. He doesn't want answers; they would collapse his metaphor and destroy his theme. As countless love songs have told us, falling in love with the Moon is the quintessence of unrequited, immortal desire.

Poets need to keep that desire alive to sustain the subjects and technicalities of their craft. Remove that distance and poetry leaves with it. Set foot upon the Moon and perhaps some of its allure disappears too? As Philip J. Fry from Matt Groening's animated sitcom *Futurama* (1999–2013) puts it: 'The moon was like this awesome, romantic, mysterious thing, hanging up there in the sky where you could never reach it. No matter how much you wanted to. But you're right, it's just this big dull rock.'[2]

Shortly after the global fever of 1969's Moon landing, interest waned: lunar missions were eventually phased out and the whole enterprise lost its media pulling power. This is mentioned in the film *Apollo 13*: 'The network's dumped us – one of them said that going to the Moon is about as exciting as going to Pittsburgh'.[3] Of course, going to Pittsburgh was once very exciting for some.

But perhaps this view overstates the importance of arrival. Maybe it's about the journey after all? En route to the Moon, the distance shrinking all the time, the metaphor is intensified, reducing like a good sauce. Perhaps it is stimulated and sharpened by the risk of annihilation: the vertigo of fearing yet wishing to fall?

With epic journeys in mind, another poetic Moon appears: the Moon in John Milton's *Paradise Lost* (1667), picked out by archangel Uriel who unwittingly steers Satan towards Paradise at the end of the fiend's long flight from Pandemonium:

> *Look downward on that globe, whose hither side*
> *With light from hence, though but reflected, shines;*
> *That place is earth the seat of Man, that light*

SIR P. S. HIS ASTROPHEL AND *STELLA*.

Wherein the excellence of ſweete Poeſie is concluded.

Well in the Ring there is the Ruby ſett,
Where comly shape, & vertue both are mett.

Φύσει βασιλικὸν τι κάλλος ἐστὶ, ἄλλως τε καὶ τῷ μετ' αἰδοῦς, καὶ σωφροσύνης κέκτηταί τις αὐτό. Xenoph. Sympoſ.

At London,
Printed for Thomas Newman.
Anno Domini. 1591.

Title page of Sir Philip Sidney's Astrophel and Stella (1591)

His day, which else, as th'other hemisphere,
Night would invade, but there the neighbouring moon
(So call that opposite fair star) her aid
Timely interposes, and her monthly round
Still ending, still renewing, through mid-heav'n,
With borrowed light her countenance triform
Hence fills and empties to enlighten th'earth,
And in her pale dominion checks the night. [4]

It is in character for an archangel to see the good in things (indeed Uriel doesn't suspect Satan of any ill intent) and the benign character of the Moon here is part of its appeal (and its genuine role in helping to maintain the conditions necessary for life on Earth) but exploration has taught us that its environment is lethal, life-annihilating, and the spacecraft and spacesuits that were built for the Apollo missions were scrupulously designed to transport pockets of liveable space with them to keep the crews alive. Harry Man explores 'space couture' in his collection *Lift* (2013), which features interesting poems about spacesuits.

Most fascinating about Apollo 11 is the intricacies of the journey, not the endpoint (there are two 'endpoints': getting to the Moon and, more crucially perhaps, getting back home). And it's not only the journey that strikes me but the extraordinary vehicle(s) and their relationship with gravity, orbits, escape velocities and the individual stages' moments of self-sacrifice: the pre-set rules of physics obeyed and exploited to dazzling effect.

Those breathtaking vehicles: three hulls of the immense Saturn V rocket; the *Eagle* lunar module with its descent and ascent stages; the intermediary service module piloted by Michael Collins that dropped *Eagle* off, passed around the far side of the Moon, and picked it up again like it was on a school run; the tiny metallic womb of the command module that splashed down into the ocean, utilising gravity and parachutes and coming home in a show of triumph and humility.

Somehow these vehicles seem to trace an evolutionary path from the peak of technological advancement back to a primordial crashing into the planet, into the seas, all the way back to the beginnings of life (with a wave to Pieter Brueghel the Elder and the fall of Icarus as they go). Saturn V's puny but indefatigable seed seeks out the egg of the Moon and, after union, the combined and changed travellers crash back into the amniotic waters of Mother Earth and humankind is born again. Or merely deflated once more, realising that desire has no limits, that nothing is fast or impressive enough to escape the earthy bog of self-deprecation and disillusionment.

It's easy to get tangled up in metaphor. In the technicalities of poetry, there are useful terms, coined by I.A. Richards, to describe its working parts. A metaphor has a *vehicle* and a *tenor*. The tenor is the concept, object or person meant and the vehicle is the image that bears the weight of the comparison. In 'the Moon's a balloon', the Moon is the tenor and the balloon is the vehicle.

In fiction, there have been a few attempts to reach the Moon by means of balloon, notably Edgar Allen Poe's short story 'The Unparalleled Adventure of One Hans Pfaall' (1835) and Josef von Báky's 1943 film *Münchhausen*. Fiction and prose, rather than poetry, seem to have been interested in getting to – getting onto – the Moon and cracking on with a good story: often an allegorical or prophetic one. Poets tend to hold back, maintaining the suspense at the heart of metaphor: adoring the Moon and speculating on it from afar, perhaps thinking a little too much about what the Moon can do for *them*, not what they can do for the Moon, to paraphrase JFK, the presiding spirit of Apollo 11.

New York poet Liane Strauss is contemporary poetry's most ardent selenophile. In 2015 she published *All the Ways You Still Remind Me of the Moon*, 24 poems about the Moon partly written in response to a 'rule' laid down by a revered poetry 'tastemaker' who decreed that 'You can only mention the Moon once in a collection'. Being something of an iconoclast and a mischief-maker, Liane mentions the Moon 13 times in the first poem alone and the collection delights in mining its metaphoric possibilities: all those vehicles orbiting the tenor but never quite landing:

What is the Moon?
The moon is not a box, or a terracotta pot,
though it spins like clay at the heart of a wheel
and I've packed it full of our most precious possessions
since it always comes with us when it's time to go.
[...]
The moon is not a toy, or a white pill,
though it skitters like a top winding down,
rolls, bounces, bounds, hides, pops, loves to be found—
and I have swallowed so many! [5]

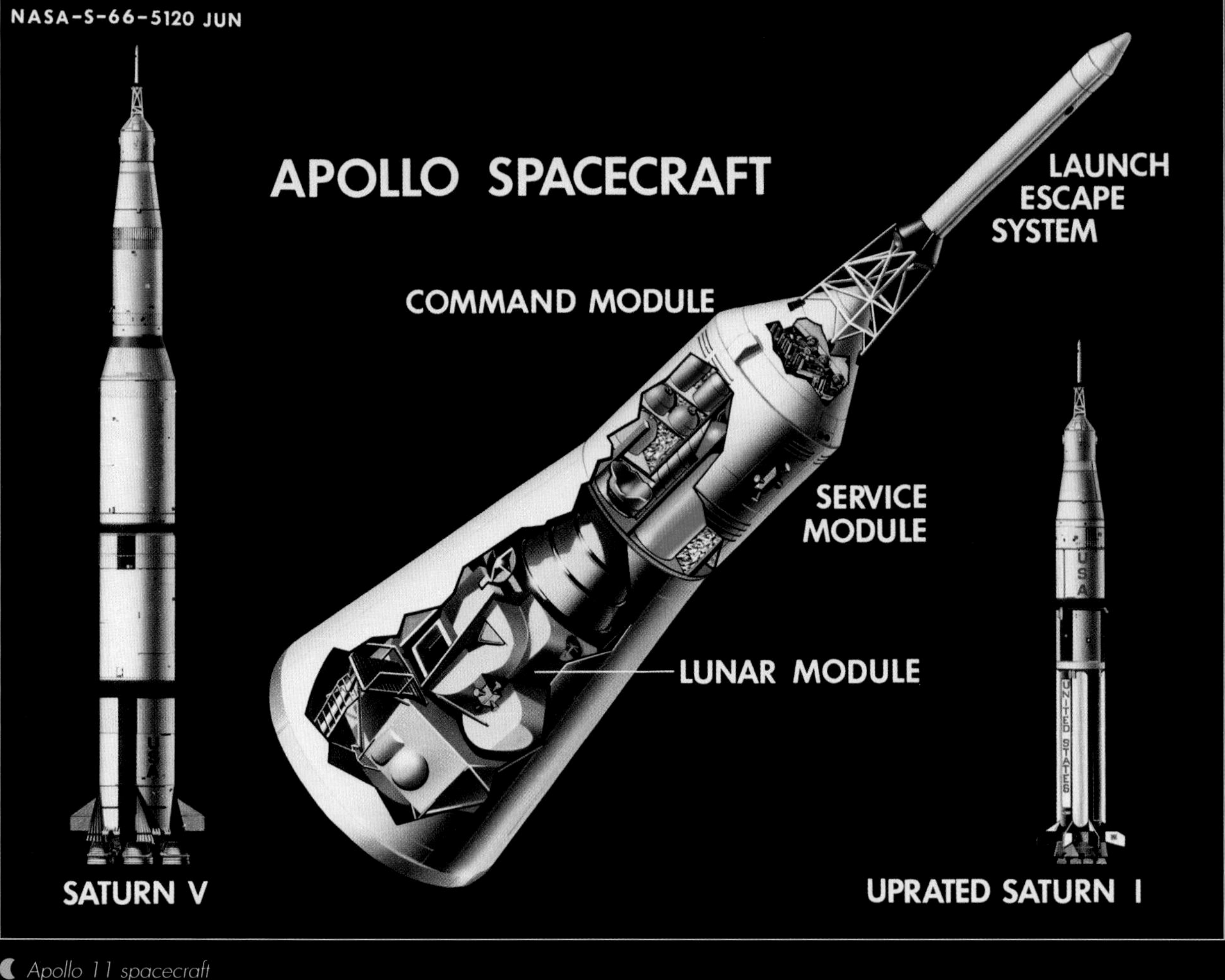

Apollo 11 spacecraft

Rebecca Elson, the late astronomer-poet, takes the absence and the elusiveness yet further with her poem, 'What if there Were No Moon?' (1999), here quoted in full:

> *There would be no months*
> *A still sea*
> *No spring tides*
> *No bright nights*
> *Occultations of the stars*
> *No face*
> *No moon songs*
> *Terror of eclipse*
> *No place to stand*
> *And watch the Earth rise.*[6]

Here mere distance turns to pure loss, even while the poet does, in this case, stand on the Moon. She looks back at the Earth with its time, its tides, its night skies, our songs and poems; she looks back at us, at the planet upon which we stand and yet, in so many ways, neglect and abuse. Perhaps we're not distant enough from Earth to appreciate it properly?

I mentioned how interest in lunar exploration waned quickly after the triumph of 1969 and the bathos of planting our feet on that once tantalising metaphor. It makes sense that, soon after, a new phase of the Moon entered popular culture with the TV series *Space: 1999*. First broadcast in 1975, the opening episode 'Breakaway' involves the Moon being blasted out of Earth's orbit by a nuclear catastrophe. The two series of *Space: 1999* depict the Odyssean wanderings of the Moon and the surviving Alphans (human

Space: 1999 *(1975)*

denizens of Moonbase Alpha) as they search fruitlessly for a new home with a safer environment, richer resources and the chance to resettle humanity.

What's the first thing the Moon does after we've 'conquered' it? It leaves us. After we have sated our desire by attaining the distant object, what do we do? We push it away again, to recreate the desire we are sorry to have lost.

Of course, the Moon may have other plans that don't involve us. Through less explosive means than nuclear cataclysm, the Moon *is* leaving us. Every year it moves away from the Earth and out of its current orbit by about 4 cm.[7] I've read that this is the rate at which our fingernails grow, with their half-moon cuticles.

The fact remains that no recognised poet has actually stood on the Moon (Apollo 15 astronaut Al Worden did publish a volume of poetry but he stayed in orbit on the command module) and, even in the imminent era of Virgin Galactic and SpaceX, very few people are likely to do so again. In the UK, few can afford the train fare from London to Glasgow unless the ticket is booked months in advance. At least on *Eagle* and in the command module you were guaranteed a seat.

So, we continue to delight in the metaphorical richness of our nearest celestial neighbour; that bright sailor with its hand on our tiller, that deep well of poetry. And it is a deep well. The American blogger Tee Zehan posted an article about Moon metaphors that dips a toe into the vast ocean of possible comparisons and purposes. Three ones of my own:

The rolling eye of a first-calf heifer.
The wakeful night light in Lazarus's dream.
A coin you must never exchange for goods.

For Moon lovers in search of poetry there's a wealth of fascinating and vital poems and poets to discover. May you never reach your destination but stay safe on the trip.

The Science Fiction of Edgar Allan Poe *(1976)*

 Whose Moon? The Artists and the Moon

Melanie Vandenbrouck

In Farideh Lashai's *Catching the Moon* (2012), a rabbit sees the Moon's reflection in a well and jumps in to catch it. Moon and rabbit become one. Lashai's installation piece reminds us that the Moon is a mirror for our dreams as much as for ourselves. While public interest in lunar travel waned as Apollo 17 Commander Eugene Cernan stepped off its dusty plains in 1972, artists have continued to look up, reflecting on the legacy of the Apollo era, interrogating the future of space exploration, and what the Moon means for humankind.

Although the sensory experience of being on the Moon remains, to date, limited to 11 test pilots and a scientist, artists have continually tried to convey its sight, smell, sound and sense of space for us all. Three decades after the historic Moon landing, photographic artist Michael Light dived into NASA's Apollo archive of 32,000 master negatives, resuscitating iconic shots and unseen images, distilled to 128 images for the exhibition and book *Full Moon* (1999). The Apollo astronauts spoke of the 'velvety blackness' of space, and Light devised a new black ink, Lunar Nero, to render the otherworldly darkness of the lunar sky against the bright lunarscape, producing strikingly luminous visions of the Moon that seemed fresh, yet all the more familiar.

In 1972, Apollo 16 astronaut Charles Duke described the smell of the Moon, which lingered on the regolith impregnating the astronauts' equipment, as like that of gunpowder. Enlisting the help of 'flavourist' Steven Pearce to recreate the lunar scent, 'Moondust Natural R342', artistic duo We Colonised the Moon produced *Moon, Scratch & Sniff* (2010), a limited edition silkscreen print whose pervasively potent aroma permeates its surroundings long after it has been scratched. As Sue Corke, one half of We Colonised the Moon, puts it:

> *smell, place, and memory are very closely linked. No one who sniffs our postcard from the moon is ever likely to go there. Yet now I hope this is a smell, similar to a freshly struck match, which will always remind them of it.*[1]

Just as olfactory art appeals to memory to enhance a sense of reality, sound resonates in subjective ways. Experience designer Nelly Ben Hayoun challenged lunar scientists to imagine the sound of Neil Armstrong's boots on the Moon. This was asking the impossible – there is no noise in the vacuum of the Moon's atmosphere-less environment – yet they played along, trying to conceive the abrasive quality of sharp dust crunched beneath heavy boots. One inspired scientist pictured 'bare feet on cow pasture', the 'sound of French toast cooking in the morning on a frying pan' or a 'maple leaf falling from a tree lying in a pool of soft fresh snow'.[2] Common to all their responses is the conjuring up of familiar images to generate something inherently alien. *The Moon Dust Remix* (2012) soundtrack created by this 'Moon Chorus' merges original footage from the Apollo 11 mission with a voiceover of descriptive musings and mouthed noises.[3]

Lunar regolith has also prompted artists to think about what the Moon is made of. Samples collected during the last Apollo mission – the only mission to include a scientist, in geologist Harrison Schmitt – inspired Spencer Finch to create *Moon Dust (Apollo 17)* (2009) a suspended sculpture of 150 chandeliers and 417 incandescent bulbs, representing an accurate three-dimensional scale model of the Moon's atomic make-up. The sizes of the bulbs vary according to the chemical element they represent: the smallest represent helium, the larger ones heavier atoms, like oxygen, iron and chromium.

But what of returning to the Moon? Agnes Meyer Brandis's *Moon Goose Analogue: Lunar Migration Bird Facility* (2011) resurrected Francis Godwin's seventeenth-century tale in which the protagonist flies to the Moon in a chariot drawn by wild geese. Brandis raised 11 hatchlings into fully-fledged Moon geese, submitting them to astronautic training ahead of their own lunar mission. If not a flock of geese, who should be going to the Moon in the wake of the 12 white American men who wandered its slopes? In 1999, Aleksandra Mir reflected that 'at this point in history, it is still clear that if a woman wants to land on the Moon, she will have to build it for herself'. Her intervention had been prompted by feminist concerns, but her public

A still from A Space Exodus, *video, 5' 24", Larissa Sansour, 2009*

Cristina de Middel, Umeko, from The Afronauts, *2012*

invitation to participate in turning a Dutch beach into a landing site also took on the community spirit of a beach clean-up, and tackled diversity as well as gender inequality. Following Mir in proclaiming herself 'First Woman on the Moon' on these newly dug sandy plains, others declared themselves 'The First Black Man on the Moon', 'The First Gay Man', and 'The First German'.[4]

Questions of sovereignty and politics are played out in Larissa Sansour's A *Space Exodus* (2009), in which the artist cast herself as the first Palestinian woman on the Moon. Sansour took inspiration from the distinctive aesthetic, cinematic slow motion and dramatic close-ups of Stanley Kubrick's *2001: A Space Odyssey*. Arabesque chords lend the musical score a Middle-Eastern feel, while the Palestinian colours replace the Apollo missions' Stars and Stripes to evoke the dream of a Palestinian state. Cristina de Middel's *The Afronauts* (2012) looks at Zambia's thwarted attempts to join the Space Race at a time when the African subcontinent was going through a wave of decolonisation, even as the USA and USSR were turning their imperialist agendas towards space. In 1964, Edward Makuka, self-appointed director of Zambia's space programme, trained ten men, two cats and a seventeen-year-old girl to go to the Moon. De Middel reconstructed this story in a series of photographs, book and video work in which her own photography is mixed with archival footage. The photographs' dynamism *(Hamba)*, hope *(Umfundi)*, manifest destiny (*Iko Iko* is your archetypal 'hero shot'), makeshift conditions (*Botonguru*) either echo or contrast with the Apollo space programme. Likewise, clad in Dutch-wax spacesuits, Yinka Shonibare CBE's own space pioneers in *Space Walk Astronauts* (2002) conflate concepts of colonial expansion, gender and race with a failed mission – the male and female figures float in space around a replica of the Apollo 13 command module, renamed in homage to civil rights movement leader, Martin Luther King Jr., who was assassinated in 1968, at the height of the Apollo era.

As plans to return to the Moon gather momentum again, artists have turned to imagining life there. Tom Hammick's oneiric *Lunar Voyage* series (2017) conjures a world in which the space traveller, plunged into self-inflicted solitude, dreamily reminisces about life on Earth, and the lush world and friends left behind. Alicia Framis set up Moon Life Foundation, a collaborative platform for a 'democratic, peaceful, artistic and cultural investigation of space', engaging researchers and the public to contribute to the innovative design of a habitat on the Moon. Merging minimalist chic with space age vintage, her *Moon Life Concept Store* (2010) invites us to consider the necessary accessories for life on a hostile environment that cannot naturally harbour life.[5]

Similar democratic aspirations led Mikael Genberg to devise the *Moonhouse Project* (2014), which aimed to install a self-assembly house on the Moon: 'Putting a house on the Moon really should be impossible', mused Genberg on the project's promotional video, 'but through crowdfunding, through the Internet, through there being enough individuals going together proving this, that means that we can do anything ... [Building] a house that belongs to all of us'.[6] Having explored engineering possibilities and identified a potential site for human settlement, the requisite equipment was meant to be shipped by SpaceX in 2015. Funding did not materialise and the project was never realised. Meanwhile, robotic exploration does not capture the imagination as powerfully as human exploration – perhaps a reason why the Google LunarXPrize, which appealed to entrepreneurial spirit, was prematurely cancelled. For now, it seems that space agencies still have the better shot at launching, landing and building a settlement on the Moon. Jorge Mañes Rubio joined the European Space Agency as artist in residence in 2016 to design a *Moon Temple* for lunar colonists to bury their dead and worship. Using the same approach as Foster and Partners for their Moon Village prototype, the artist proposes to three-dimensionally print the structure with regolith. Early visualisations show a structure drawing inspiration from eighteenth-century French utopian architectural projects, with a distinctively brutalist feel, suited to the barren terrains of the Moon.

Yet while artists are turning their eyes to the future of space exploration, they too continue to reflect on the ways in which the Moon deeply affects life here on Earth. In his *Afro Lunar Lovers* (2003), Chris Ofili evokes the romance of moonlight. One should not mistake this print for innocuous romantic musing – the red, black and green evoke the flag of the Pan-African movement, and lend a militant persona to his tenderly embraced lovers. Plotted along the course of the Docklands Light Railway in London, Alison Turnbull's *Time and Tide* cycle shelters (2010) reminds us that the Moon remains our constant companion, as it affects our daily lives. One side adorned with lunar phases, the other with the hydrographic survey's tidal predictions in cities across the world, the shelters evoke the Moon's calendrical role and effects on the world's oceans.

Tom Hammick, Chamber, a woodcut print, from Lunar Voyage*, 2017*

Chris Ofili, Afro Lunar Lovers, *an embossed giclée print, 2003*

Once upon a time a man met the Moon and decided to spend the rest of his life with her. The verses of Leonid Tishkov's *Private Moon* series (2003–ongoing) tell us how the Moon 'helps us to overcome our loneliness in the universe by uniting us around it', while the poetic photographs chart his relationship with the Moon, from its discovery in his attic, to their travels around the world.[7] With Tishkov, we are reminded of our long and enduring affair with the Moon. As a species, we have been, remain, and forever will be, moonstruck.

'Like a lunar unicorn / under the covers / she shines even brighter', from the series Journey of the Private Moon, *Moscow, 2003–05 by Leonid Tishkov*

The Geopolitics of the Moon

Jill Stuart

> *We set sail on this new sea because there is new knowledge to be gained, and new rights to be won, and they must be won and used for the progress of all people. For space science, like nuclear science and all technology, has no conscience of its own. Whether it will become a force for good or ill depends on man, and only if the United States occupies a position of pre-eminence can we help decide whether this new ocean will be a sea of peace or a new terrifying theatre of war…*
>
> *There is no strife, no prejudice, no national conflict in outer space as yet. Its hazards are hostile to us all. Its conquest deserves the best of all mankind, and its opportunity for peaceful cooperation may never come again…*
>
> *We choose to go to the Moon in this decade and do the other things, not because they are easy, but because they are hard; because that goal will serve to organize and measure the best of our energies and skills, because that challenge is one that we are willing to accept, one we are unwilling to postpone, and one we intend to win, and the others, too.*
>
> John F. Kennedy, 12 September 1962

The Moon occupies two extremes in humanity's imagination: the noble desire to discover and recognise our shared fate; and the drive for political power and prestige. Exploration of space is often associated with phrases such as the 'human imperative to explore,' 'the common heritage of mankind,' and 'collective humanity'. Yet, space exploration has also been driven by geopolitics and military aspirations. Scientific aspirations have been key drivers behind the examination and penetration of outer space, yet the funding for such activities has often come from governments pursuing political agendas. The complex relationship between these two elements has led to the negotiation of intricate international laws to govern the cosmos. This chapter explores the geopolitics behind those laws in the context of lunar exploration past and present, and looking into the future.

Who owns the Moon?

How humankind has laid a claim to the Moon is imprinted on its barren surface by the objects that were left there. Few have more symbolic significance than flags.

Although most lunar scientists believe they are now bleached white by solar radiation, there are currently six American flags on the Moon, one for each Apollo mission that landed there. Interestingly, despite the apparent patriotic symbolism, the first of the six flags is actually thought to have been procured only at the last minute by NASA and bought off the shelf from a department store for little more than US$5 (although which store and which manufacturer remains the topic of some debate). The Apollo 11 astronauts also left a commemorative plaque that read, 'We came in peace for all mankind'. Ironically, the USA was engaged in fierce fighting in Vietnam at the time.

Five other entities have since left flags on the Moon: the Soviet Union; China; Japan; India; and the European Union. Of these, only the Soviet Union and China 'soft landed' spacecraft there. The other flags were remotely delivered, i.e. strategically crashed into the surface for scientific reasons (to study the impact); one could argue that each was symbolic of political interests.

Speaking of crashing into the Moon, the USA considered nuking the Moon during the Cold War rather than landing on it. Project A119, also known as 'A Study of Lunar Research Flights', assessed the scientific (but ostensibly political) benefits of sending a nuclear bomb to the Moon. The explosion would have been visible from Earth and so would have demonstrated US military might. The project was scrapped, however, because it was concluded that the American public would not support it.

Uncertainties in the era were also reflected in popular culture. In 'The Intruder', an episode of the British children's TV show, *The Clangers*, which aired on 28 December 1969, a lunar lander touches down on the Moon's surface. Although this was several months after the Apollo 11 landing, the script had been drafted well beforehand. The animators therefore

A footstep on the Moon, 21 July 1969, Apollo 11

Apollo 11 astronaut Edwin 'Buzz' Aldrin beside the US flag he planted on the Moon, 21 July 1969

The plaque left on the Moon by the Apollo 11 astronauts

AFSWC-TR-59-39

SWC
TR
59-39
Vol I

HEADQUARTERS

AIR FORCE SPECIAL WEAPONS CENTER

AIR RESEARCH AND DEVELOPMENT COMMAND

KIRTLAND AIR FORCE BASE, NEW MEXICO

A STUDY OF LUNAR RESEARCH FLIGHTS
Vol I

by

L. Reiffel

ARMOUR RESEARCH FOUNDATION

of

Illinois Institute of Technology

19 June 1959

Cover of 'A Study of Lunar Research Flights – Vol I' (1959)
a top-secret plan to detonate a nuclear weapon on the Moon

A hybrid American-Soviet flag planted on the Moon in 'The Intruder', an episode of The Clangers, *1969*

had the spaceman plant a flag that included both American and Soviet icons, as it was uncertain which superpower would land first.

When Apollo 11 astronaut Edwin 'Buzz' Aldrin planted the first American Stars and Stripes on the Moon on 21 July 1969, he performed a clearly symbolic, nationalistic act, yet he was not effectively claiming the Moon for the USA. For centuries, planting a flag into a territory has been understood as staking a territorial claim, but in this case it was readily understood that actual ownership of the lunar surface was not being implied and that the Moon remained neutral territory. This was largely because two years earlier, between 27 January and 10 October 1967, the Outer Space Treaty (OST) had been completed and widely ratified by countries around the world through the United Nations. The treaty established outer space as 'neutral territory' (*res communis*), 'the province of all mankind', to be used without discrimination and on a basis of free access, not subject to 'national appropriation by claim of sovereignty, by means of use of occupation or by any other means'.[1] In other words, outer space belonged to no-one and to everyone. This is a lovely sentiment, but it is worth going further back in time to consider how the OST came about in the first place.

Early rocket science and the Second World War

As far back as the late-fifteenth century, Leonardo da Vinci and others had dreamed of reaching the stars by combining gunpowder with rocket science. Important proposals were made in the late-nineteenth century by Russian scientist Konstantin Tsiolkovsky and others, but it wasn't until the 1920s that rocket science was developing to the point that reaching high atmosphere or Earth orbit was becoming feasible. In 1920 the Smithsonian Institute issued a press release on a grant application from the American scientist and spaceflight pioneer Robert H. Goddard (1882–1945). Couched within Goddard's richly detailed proposal, *A Method of Reaching Extreme Altitudes* (1919), was a passing comment that eventually the Moon could be reached by human-made objects, though the paper also stated that such a mission was 'not of obvious scientific importance'.[2] The American press seized on and sensationalised the claim with reactions ranging from the enthusiastic – 'the Bronx Exposition, Inc. offers the use of Starlight Amusement Park for the purpose [of hosting the rocket departure], and at the same time will be happy to provide all of the facilities needed for the occasion,'[3] to the disdainful – 'Of course he only seems to lack the knowledge ladled out daily in high schools'.[4]

Whilst Goddard's proposal seemed fantastical, and uniquely captured the public's attention, scientists in other parts of the world were working towards the same goal. In the interwar period, budding rocket scientists worked individually but also formed non-governmental collectives for the purpose of advancing and testing rocket science. For example, the All-Union Society for the Study of Interplanetary Flight was a non-military society in the USSR, and the Verein für Raumschiffahrt (Society for Space Travel) or VfR was an amateur rocket association founded in Germany in 1927, but to which other nationals were welcome to contribute. Some of these scientists had begun to collaborate across national boundaries, and even the self-isolated Goddard sent a copy of his research to the German rocket-scientist Hermann Oberth, with whom he briefly engaged in correspondence. Oberth's assistant, Wernher von Braun, was influenced by Goddard's ideas and carefully studied his papers.

However, politics soon intervened. By the late 1930s, it was apparent that the Second World War was brewing. Government and military leaders recognised something about these rockets that scientists were creating: that they could have a military application. Space launches normally consist of two main pieces – the rocket and the payload. It became apparent that rockets designed to carry benign payloads such as satellites could also carry missiles across long distances. Rocket scientists were either tempted by their governments with funding or coerced into militarising their research. In France, airplane engineer and spaceflight theorist Robert Esnault-Pelterie convinced the French War Department to fund his research into ballistic missiles as early as 1930. Other scientists were told by their governments to prepare for war, not space exploration. In Germany, von Braun was enlisted by the Nazi regime to continue his ground-breaking research at Peenemünde, where he led the team that designed and developed what would become the infamous V-2 rocket. The research of a leading light of Soviet rocketry Sergei Korolev (1907—66) was likewise militarised, but carried out from imprisonment in the gulag from 1938 until 1944. Although Goddard had proposed military applications for his research on rocketry when the US entered the First Wold War in 1917, during the Second World War it was not until the attack on Pearl Harbor that his work was militarised.

As wartime tensions simmered, the long-range missile was born. Germany's V-2 would prove to be a devastating weapon used across Europe. In 1944 it would also become the first man-made object to cross the Kármán Line, lying at an altitude of 100 kilometres above sea level, thus becoming the first object that mankind placed into 'outer space'.

Early space exploration and the Cold War

At the end of the Second World War, renewed efforts focussed on using rocketry for space exploration. Many individuals involved in these efforts were driven by the desire to explore, discover and learn about the cosmos – yet power politics were still playing a driving role. A few days before Germany's surrender, von Braun and his team of specialist engineers handed themselves over to US forces and were transferred to America. The Peenemünde facilities were scoured for clues about the technology developed there, first by the USA, then by the USSR, who released Korolev from the gulag to piece together what had been left by the Americans. By 1947, the geopolitical map had been redrawn, with two superpowers, the Soviet Union and United States, now locked in the Cold War, characterised by mounting tensions between the Eastern bloc (the USSR and its satellite countries) and the West (the USA and NATO allies). As both superpowers were equipped with nuclear weapons, the threat of mutually assured destruction was real: if one country were to launch its nuclear weapons and the other responded, it could result in global devastation. Each superpower thus sought to demonstrate its capabilities and strengths without resorting to direct confrontation. Both von Braun and

Dr. Robert H. Goddard and a liquid oxygen-gasoline rocket in the frame from which it was fired on March 16 1926, in Auburn, Massachusetts

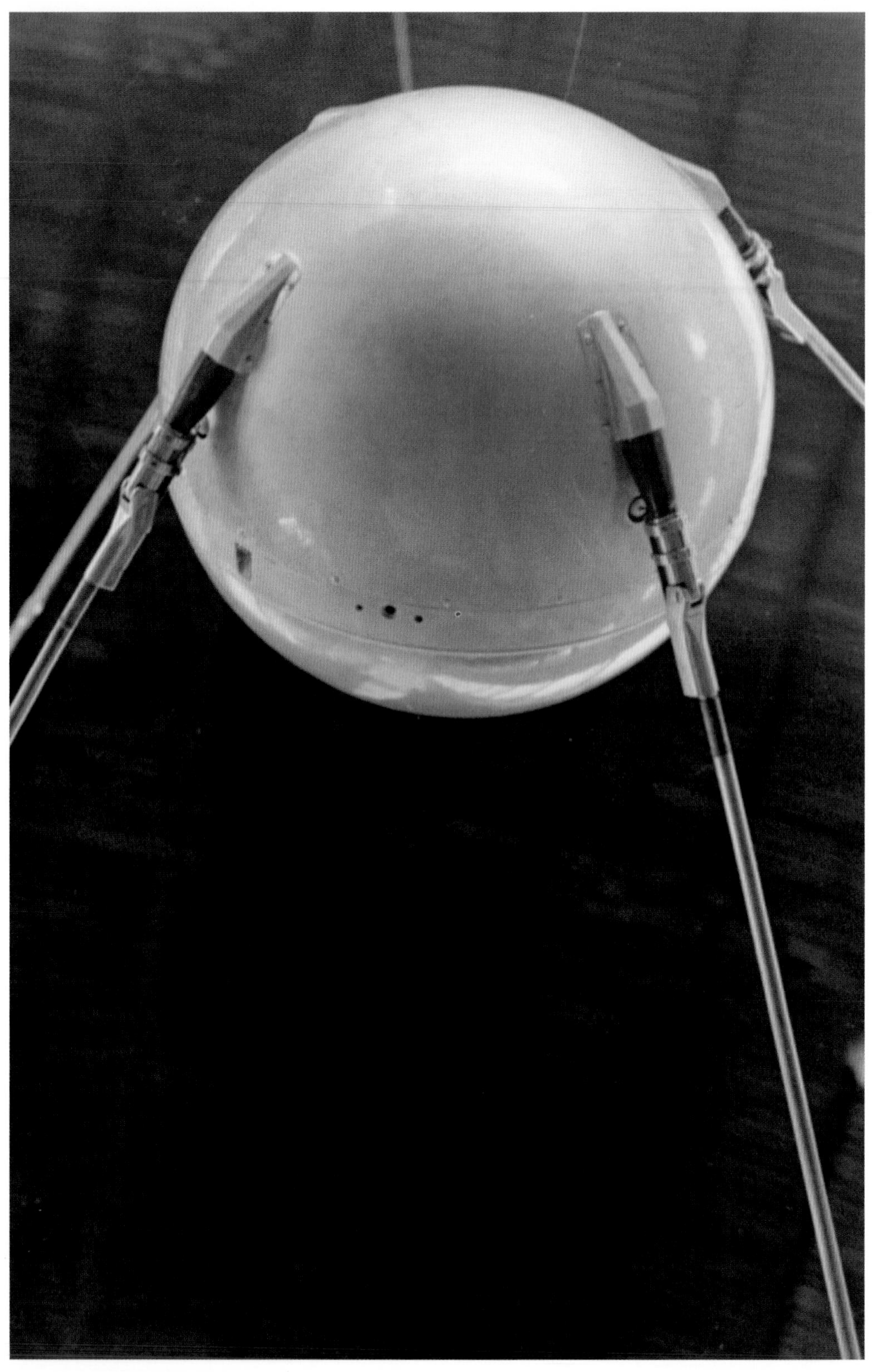

A close-up view of the full-scale mock-up of the Sputnik 1 spacecraft on display at the Soviet Pavilion at the Paris Air Show, France

Korolev took the opportunity to carry out their research into space rocketry, which would lead to the American and Soviet space programmes' greatest achievements.

The Outer Space Treaty of 1967

Pursuing activities in space provided a unique opportunity for the US and USSR to compete with each other in the Cold War context. They could rattle their sabres through space programmes in a way that did not require overt military activity but presented a useful display of technological, economic, ideological and cultural might. Space activity requires strong technological capabilities, which can be turned into military force-multipliers, particularly in the case of rockets. It is expensive and thus potentially indicative that a country has a strong economy. As such, in order to counter the repeated blows of the USSR's irresistible advance and series of 'firsts' in outer space, increased funding was injected into the Apollo space programme (1960–72) that followed Project Mercury (1958–63) and Project Gemini (1961–66) as the US's human spaceflight programmes. At its peak in 1966, Apollo cost about 0.8 per cent of US GDP and employed some 400,000 personnel, while NASA's budget consumed 4 per cent of federal spending in 1964–66. What is more, to have a space programme shows rivals that a country is politically united and strong. It puts the nation at the forefront of human exploration in a way that demonstrates global prestige. In addition, the Soviet Union and USA could work with amenable partner countries on space activities in a way that shored up alliance chains and established international leadership in the bipolar international context.

In this bubbling cauldron of geopolitical competition, where it was apparent that technological developments were likely to make the penetration of space imminent, both the USSR and the USA went to the United Nations (UN) to negotiate a treaty to govern activity in the cosmos. The UN first focused on outer space issues shortly after the USSR's launch of *Sputnik 1* in 1957. The UN Committee on the Peaceful Uses of Outer Space (UNCOPUOS) was created in 1958 (and still meets regularly in Vienna), and the OST was eventually ready for ratification in 1967.

Given the overwhelming preponderance of power of the USA and USSR during this era, it is curious that the two global hegemons (the states with pre-eminent economic, political and military control within the international system) sought to negotiate a treaty in the first place: why not leave it as an anarchic realm, a playground for those rare countries who had the resources and technology to develop space capabilities? The wording of the OST uses altruistic language relating to space as 'the province of all mankind', yet the motivations behind its drafting were partly strategic. The two Cold War superpowers knew that they may not be able to physically or technologically dominate outer space unilaterally, so the best option for each was to ensure that their rival couldn't either. As such, both wanted to proceed with caution by agreeing to mutual self-restraint. Still, this left open the very big question of what sort of restraints they would agree to. What should a treaty governing outer space consist of? Previous developments in international law negotiated through the UN offered some possible analogies as a starting point. Two legal regimes were particularly relevant, those concerning airspace and the high seas.

Norms regarding the regulation of airspace had evolved for centuries, from Roman times when ownership above land was granted to the owner of the *terra firma* beneath it, to eighteenth-century debates about the legality of flying hot air balloons. Commercial aviation in the twentieth century spurred on formalisation of airspace governance, with the International Civil Aviation Agency created as a specialist branch of the UN in 1944. Subsequent treaties and agreements established that governments had sovereign authority over the vertical area above their territories, with passage of objects allowed with permission. Thus, if outer space were to be modelled on an airspace analogy, the territory above a country would extend into outer space and that country would maintain sovereign governance over that space. A foreign country's objects could pass through such territory, but only with the oversight of the governing state below.

Mare liberum translates as 'freedom of the seas' and since the sixteenth century it has been a guiding principle in the governance of the world's oceans. While 'unobstructed passage' has been in practice for centuries, it was formally codified in the United Nations Convention on the Law of the Seas in the Fifties and early Sixties. This law dictates that beyond a country's territorial waters (12 miles from their coast) the oceans are neutral territory. Objects on the high seas (such as ships) must be registered with an individual country, but are given free passage on the oceans. The oceans transcend international boundaries and are *terra nullis*. Objects placed there are the responsibility of the country under whose flag they sail. Thus, if outer space were to be modelled on the high seas analogy, a country's airspace would end beyond a certain point and outer space would begin. Objects placed there would be the responsibility of the launching state but would otherwise have free passage through the vacuum of space.

The USA and USSR disagreed on which model should be used to draft a treaty for the cosmos. The Soviets initially wanted an airspace analogy, which would extend sovereignty into outer space and restrict overflight of future satellites above a country's territory. The Americans pushed for the high seas analogy, which would give objects in space free passage. The difference in the two superpowers' positions was mainly strategic. In part, logic tells us, this related to the fact that the Soviet Union was a larger territory and hence the airspace analogy would grant them a larger swathe of space territory. However, the primary motivator behind their respective interests related to reconnaissance. The closed Soviet political system greatly limited American information on their enemy, and hence the high seas model and free passage would allow for greater reconnaissance opportunities from orbit. The Soviets had an advantage over the USA in information gathering because the American political and media system was more open. As such, it was in the interest of the Soviets to support an airspace analogy and the control it would grant over objects in orbit over their territory.

And yet the OST that was agreed upon dictates free passage and neutral territory, following the high seas analogy. Indeed, the USSR's interests eventually changed, for two main reasons. Firstly, by the time the OST was in

"All the News That's Fit to Print"

The New York Times.

LATE CITY EDITION
U. S. Weather Bureau Report (Page 33) forecasts: Cloudy and cool today and tonight. Mostly fair tomorrow.
Temp. range: 65—53. Yesterday: 62.4—49.2.

VOL. CVII..No. 36,414. © 1957, by The New York Times Company. Times Square, New York 36, N. Y. NEW YORK, SATURDAY, OCTOBER 5, 1957. 10c beyond 100-mile zone from New York City FIVE CENTS

SOVIET FIRES EARTH SATELLITE INTO SPACE; IT IS CIRCLING THE GLOBE AT 18,000 M. P. H.; SPHERE TRACKED IN 4 CROSSINGS OVER U. S.

HOFFA IS ELECTED TEAMSTERS' HEAD; WARNS OF BATTLE

Defeats Two Foes 3 to 1 —Says Union Will Fight 'With Every Ounce'

Text of the Hoffa address is printed on Page 6.

By A. H. RASKIN
Special to The New York Times.

MIAMI BEACH, Oct. 4—The scandal-scarred International Brotherhood of Teamsters elected James R. Hoffa as its president today.

He won by a margin of nearly 3 to 1 over the combined vote of two rivals who campaigned on pledges to clean up the nation's biggest union.

Senate rackets investigators and Hoffa critics in the union rank-and-file immediately opened actions to strip the 44-year-old former warehouseman from Detroit of his election victory.

A jubilant Hoffa exhibited, however, greater concern over the possibility that his union might be ousted from the American Federation of Labor and Congress of Industrial Organizations. He appealed for time to prove that he could make the teamsters "a model of trade unionism."

The parent organization has ordered the 1,400,000-member Teamsters Union to get rid of corrupt leadership by Oct. 24 or face suspension. Hoffa said he felt actions by the union at its week-long convention here should satisfy the federation.

Associated Press Wirephoto
IN TOKEN OF VICTORY: Dave Beck, retiring head of the Teamsters Union, raises hand of James R. Hoffa upon his election as union's president. At right is Mrs. Hoffa.

FAUBUS COMPARES HIS STAND TO LEE'S

Says He Will Remain Loyal to People of Arkansas— All Is Quiet at School

By HOMER BIGART
Special to The New York Times.
LITTLE ROCK, Ark., Oct. 4

Flu Widens in City; 10% Rate Predicted; 200,000 Pupils Out

By ROBERT ALDEN

Asian influenza continued to spread through the city yesterday.

Commissioner of Hospitals Morris A. Jacobs reported that there were ten times more respiratory infections than during the comparable period a year

ARGENTINA TAKES EMERGENCY STEPS

State of Siege Proclaimed in Buenos Aires Region —Arrests Reported

By Reuters.
BUENOS AIRES, Oct. 4—A state of siege, suspending

COURSE RECORDED

Navy Picks Up Radio Signals—4 Report Sighting Device

By WALTER SULLIVAN
Special to The New York Times

WASHINGTON, Saturday, Oct. 5—The Naval Research Laboratory announced early today that it had recorded four crossings of the Soviet earth satellite over the United States.

It said that one had passed near Washington. Two crossings were farther to the west. The location of the fourth was not made available immediately.

It added that tracking would be continued in an attempt to pin down the orbit sufficiently to obtain scientific information of the type sought in the International Geophysical Year.

[Four visual sightings, one of which was in conjunction with a radio contact, were reported by early Saturday morning. Two sightings were made at Columbus, Ohio, and one each from Terre Haute, Ind., and Whittier, Calif.]

Press Reports Noted

Soviet newspapers reported several weeks ago that the Soviet satellites would broadcast on frequencies in the neighborhood of twenty and forty megacycles. More exact frequencies were given by Soviet scientists at a conference on rockets and satellites that took place here this week.

Presumably the Naval Research Laboratory, which is responsible for the United States satellite program under the National Academy of Sciences, immediately set up receivers on those frequencies.

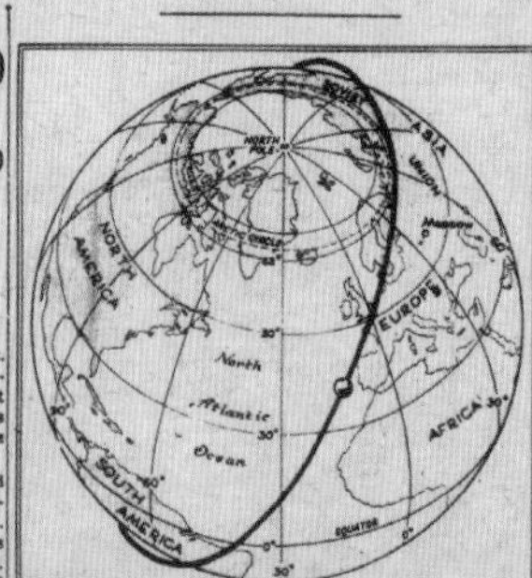

The New York Times Oct. 5, 1957
The approximate orbit of the Russian earth satellite is shown by black line. The rotation of the earth will bring the United States under the orbit of Soviet-made moon.

Device Is 8 Times Heavier Than One Planned by U.S.

Special to The New York Times.

WASHINGTON, Oct. 4—Leaders of the United States earth satellite program were astonished tonight to learn that the Soviet Union had launched a satellite eight times heavier than that contemplated by this country.

Dr. Joseph Kaplan, chairman of the United States program for the International Geophysical Year, described the 184-pound weight as "fantastic." The heaviest American satellites are to weigh twenty-one and a half pounds.

The actual launching, nevertheless, did not take the American scientists by surprise. At the end of working sessions on

SATELLITE SIGNAL BROADCAST HERE

Impulse Carried on Radio and TV—First Reported by Long Island Station

560 MILES HIGH

Visible With Simple Binoculars, Moscow Statement Says

Text of Tass announcement appears on Page 3.

By WILLIAM J. JORDEN
Special to The New York Times.

MOSCOW, Saturday, Oct. 5—The Soviet Union announced this morning that it successfully launched a man-made earth satellite into space yesterday.

The Russians calculated the satellite's orbit at a maximum of 560 miles above the earth and its speed at 18,000 miles an hour.

The official Soviet news agency Tass said the artificial moon, with a diameter of twenty-two inches and a weight of 184 pounds, was circling the earth once every hour and thirty-five minutes. This means more than fifteen times a day.

Two radio transmitters, Tass said, are sending signals continuously on frequencies of 20.005 and 40.002 megacycles. These signals were said to be strong enough to be picked up by amateur radio operators. The trajectory of the satellite is being tracked by numerous scientific stations.

Due Over Moscow Today

Tass said the satellite was moving at an angle of 65 degrees to the equatorial plane and would pass over the Moscow area twice today.

"Its flight," the announcement added, "will be observed in the rays of the rising and setting sun with the aid of the simplest optical instruments,

The New York Times *front cover, 5 October 1957*

negotiation, the USSR had launched Earth's first manmade satellite, *Sputnik 1*. On 5 October 1957, as *The New York Times* put it, '[T]he "new Soviet society"... turned the boldest dreams of mankind into reality'.[5] *Sputnik* had a 'Space Pearl Harbor effect'. As a direct response to *Sputnik*, then-President Dwight D. Eisenhower founded the National Aeronautics and Space Administration (NASA) in 1958.

Sputnik 1 was an international political victory for the USSR and a boost for domestic national pride, but also had implications for the future governance of space. The satellite circled the Earth in low orbit, passing overhead every 98 minutes and emitting a beep that could be heard by anyone on Earth in possession of capable radio technology; later President Lyndon B. Johnson referred to it as the 'beep heard around the world'.[6] In passing over various countries on its journey, *Sputnik 1* established the norm of free passage for outer space. If the Soviets continued to argue for extended sovereignty into outer space based on the airspace analogy, they had already undermined their own principle.

The second reason reflected significant geopolitical developments on Earth. The Soviet Union had previously been allied with the People's Republic of China, but in the Fifties and Sixties relations between the two deteriorated, leading to the Sino-Soviet split (1956–66). The USSR became increasingly amenable to having satellites legally allowed to pass freely through space for various uses, including spying.

The evolution of outer space law

Today the OST of 1967 has been ratified by 107 countries and signed by an additional 23. It is sometimes referred to as 'The Constitution for Outer Space'. Three other major treaties were subsequently drafted through the UNCOPUOS and widely ratified. These relate to the rescue of astronauts, liability for damage caused by objects in space – an increasingly pressing issue as space debris is building exponentially with the pace of new satellite launches – and registration of objects launched into space.

A fifth treaty drafted in 1979 relates specifically to the Moon. However, it is considered in international legal terms to have 'failed' as only 18 countries have ratified it, none of which has crewed space capabilities. The intention of the Moon Treaty was to reiterate many of the core tenets of the OST. These included banning the alteration of any celestial environment and banning ownership of celestial bodies by any organisation or person.

Why did the Moon Treaty fail? Was it so different from the Outer Space Treaty of 1967 in what it required for lunar territory? Arguably, the Moon Treaty's failure was not about the content of the Treaty itself, but rather the way space law and governance had evolved to fit the contemporary international system. By the late Seventies, there was a shift in the overall approach to the governance of outer space. In particular, with increasing numbers of countries developing space capabilities, negotiating all-encompassing treaties with large numbers of signatories was becoming increasingly challenging.

The Moon Treaty of 1975 was the last of five major outer space treaties. An additional complication by that

Timeline and United Nations treaties governing outer space

Date	
1918	End of the First World War
1920s	Developments in rocket science
1939–45	Second World War – rocketry used for military purposes; development of the V-2
1947	Onset of the Cold War
1957	Soviet Union launches *Sputnik 1*
1967	Outer Space Treaty (Treaty on Principles Governing the Activities of States in the Exploration and Use of Outer Space, including the Moon and Other Celestial Bodies)
1968	Rescue Agreement (Agreement on the Rescue of Astronauts, the Return of Astronauts and the Return of Objects Launched into Outer Space)
1969	Apollo 11 lands on the Moon
1972	Last Apollo mission leaves the Moon
1972	Liability Convention (Convention on International Liability for Damage Caused by Space Objects)
1976	Registration Convention (Convention on Registration of Objects Launched into Outer Space)
1979	Moon Agreement (Agreement Governing the Activities of States on the Moon and Other Celestial Bodies)

time was that outer space was no longer the exclusive domain of countries, even though outer space treaties were still being written in state-centric language with state signatories. As such, supplementing the four main treaties with memoranda of understandings and multilateral agreements augmented by domestic laws became a more realistic approach to the governance of outer space in the contemporary era. The failure of the Moon Treaty reflected that transition.

Looking forward, looking out

Topics such as tourism, mining colonisation and celestial ownership have legal, ethical and philosophical implications that variously capture the attention of politicians, lawyers, academics and the media. However, at the core of them all is the changing profile of those who are participating in space activities. Once primarily the domain of an elite number of states, today space is accessible to a large number of countries and a diversity of non-state entities.

This has not been a sudden revolution but rather a gradual opening up of space over more than half a century. What is notable about the past decade, however, is the degree to which there is a diversity of entities invested in space, including everyday citizens increasingly aware of our collective utilisation of space infrastructure on a daily basis (see the table below).

Consider the following comparisons:

Past	Today
In the Seventies, telecommunications consortium projects such as INTELSAT gave an expanding number of countries nascent space capabilities, albeit dependent upon 'lead' countries such as the United States.	There are 70 government and intergovernmental space agencies, 13 with launch capabilities.
The commercial space industry has been expanding since the Seventies and Eighties. Growth in this era was boosted by the popularity of products such as satellite TV and telecommunications. In addition, in the Seventies and Eighties several governments proactively pushed for greater commercialisation; for example, the US through legislation under presidents Richard Nixon and Ronald Reagan; in the late Seventies, China's Deng Xiaoping promoted space for economic goals; in the Eighties, the Soviet Union began marketing private space launches.	In the widely publicised UK Space Innovation and Growth Strategy (2015 update), it was estimated that by 2030 the global space market would be worth £400 billion. Commercial space activities range from global positioning in mobile phones to remote sensing services and use in private banking transactions ... and it is big business.
The first official space tourist was Dennis Tito, who in 2001 paid an estimated US$20 million to travel to the International Space Station.	A few competing companies are potentially on the brink of providing sub-orbital 'adventure tourism'. Tickets with Virgin Galactic are currently selling for around US$200,000 per seat.

Compared with the days of the Cold War Space Race when two main superpowers dominated space activity, the three examples above highlight how a wider number of entities are now active in space: more countries, more commercial entities and more private citizens. However, for political and legal purposes a further complication is what could be termed a 'diversification of actors,' in that these entities increasingly combine in complicated ways. Commercial companies may be partly funded by governments; government space activities may be subcontracted by private companies; individuals may purchase tickets to visit the intergovernmental International Space Station; a privately owned satellite may be launched from a ship that flies the flag of a particular country but is anchored in the neutral territory of the high seas. These sorts of partnerships are not new, but globalisation, technological developments and human ingenuity mean that space activity is being undertaken in creative ways by collaborations of entities that weren't as common in the past.

The multinational corporation Sea Launch is an example of one such entity through which different companies and governments are pursuing space activity. Sea Launch puts objects into outer space – mainly telecommunications satellites from private companies. The rockets they use come from several countries: some are military hardware originally intended for rockets built during the Cold War. The payloads are shipped and then prepared on the high seas (which are neutral territory according to international law). The objects are then launched from large ships into outer space. The ships from which the objects are launched fly 'flags' to denote their nationality. As such, this company both embodies and challenges the complicated nature of what the 'launching state' and 'sovereignty' mean ... both in outer space and also on a globalised earth. SpaceX was the first private American company to send a resupply vehicle to the International Space Station. It may be a private company, but gains considerable funding as a subcontractor from the government agency NASA. South African Denis Tito was the first person to come onto the International Space Station as a 'tourist'. He was on board the station as a 'guest' of a private tourism company. In that context, he filled a complicated role as a citizen of a country, a client of a company, someone's guest within a foreign enclave and as someone in neutral territory.

These collaborations aren't a bad thing. However, it does potentially complicate matters when it comes to outer space law and to philosophical and ethical questions regarding space politics and governance. First and foremost is the problem that the four main treaties that govern outer space are state-centric. As mentioned above, the OST pronounces that outer space is not subject to '*national* appropriation by claim of sovereignty' (my italics). In subsequent treaties, liability is assigned to a space object's 'launching state'; astronauts who land on Earth, for instance, are to be returned to their home country. These treaties were conceived at a time when activity in space was state-led and can make its application today awkward.

Consider some examples (hypothetical and actual) to illustrate the point:

- According to the OST, no *country* can lay claim to a celestial body. Does that mean that an individual or a company can? Some have made this claim, sometimes seriously and sometimes spuriously. For a period of time it was fashionable to 'buy a plot of land on the Moon' as a novelty gift, with companies involved saying they had claimed the territory there as they were not subject to the OST's 'national appropriation' or 'sovereignty' clauses. (At the risk of disappointing any proud owners, claims to land would be very unlikely to be upheld under international law and in any case several companies claimed the same plots of land many times over.)
- If an entity – perhaps a public-private partnership in the form of a company backed by government funding – extracts lunar resources such as Helium 3, which is then used to return a mission to Earth, does that resource cease to be a celestial resource and thus become free for appropriation?
- If an individual astronaut is flying as a tourist and launches from one country on a ticket bought from a tourism company based in another country, yet holds a passport from somewhere else, what 'nationality' do they have in outer space? Whose laws do they fall under on the Moon if they commit a crime or get married?

The above highlights what the potential legal issues may be for future exploration of outer space and the Moon. But to conclude on a philosophical note and a challenge to the reader, these issues raise deeper questions, not only of who represents us in space, but also who we *want* to represent us in space. Moving forward, do we want to see countries continue to dominate? It has been nearly half a century since humans last walked on the Moon, but there is renewed interest in returning. From which nation will the next flag planted by a human being on the Moon come? If not a nation, who do we want to represent us there: an international conglomerate, a multinational corporation, an individual, even? Do we care if they are there to mine its resources, and what if the purpose of doing so is to sustain a mission to other planets? There are no easy answers, but it is important to ponder the questions.

EPILOGUE

Plato
Fontenelle
Timaeus
Birmingham
Epigenes
Anaxagoras
Philolaus

We Came All This Way

Melanie Vandenbrouck

> *Now I know why I'm here: not for a close look at*
> *the Moon,*
> *but to look back*
> *at our home*
> *the Earth*
>
> Al Worden, Perspectives.[1]

To go to the Moon has been an aspiration, dream and obsession since humankind first imagined it as a world. And to go to the Moon – whether to orbit its world or step on its surface – changes you. Between 1968 and 1972, 24 men came close to its surface, 12 of which walked on it. These were certainly not sentimental men; apart from geologist and geophysicist Harrison Schmitt, they were military fighter or test pilots. And yet they returned touched by an unexpected grace. One became a poet, another an artist. Some found god, others joined counterculture. Astronaut-turned-poet Al Worden found 'meaning'. Apollo 11 Command Module Pilot Michael Collins spoke of the extraordinary peace he felt as he circled the Moon on his own while his two companions were making history on the grey world below. On every orbit, he was cut off from the whole of humankind as communications broke each time he flew behind the Moon. It was solitude as unfathomable as it was complete.

These men had seen combat or risked their lives flying prototype aircraft. They had signed up for risky, historic journeys of exploration into the unknown. Recognised as having 'the right stuff', they had sailed through a superhuman training regime in which they were subjected to physical extremes, from the amplified g-force during take-off and splashdown, to weightlessness in space. They were recognised for their sang-froid in the face of any eventuality – a cool-headedness that was to be tested during Apollo 13's desperate, near-fatal flight home. Still, nothing could have prepared them for the shock of the otherworldly sight of the Moon. Neil Armstrong had carefully rehearsed his first words as he gingerly crossed the next frontier of human exploration, making 'one small step for (a) man, one giant leap for mankind'. What he could not rehearse was his reaction to the scene around him: 'it has a stark beauty all its own', he shared with Mission Control in Houston as he surveyed land around him.[2] Armstrong was reputedly a terse man, making the simplicity and spontaneity of these words, vented as if in disbelief, all the more poignant. His more eloquent companion, fellow 'moonwalker' Buzz Aldrin, spoke of the 'magnificent desolation' that lay before him, words that have come to define the Moon ever since. This world was distinctly alien, tremendously hostile – a leak in a spacesuit or on the flimsy lunar module would have been fatal. Some of the men who followed could not resist leaving a bit of home on its soil, as if to make it more hospitable. Charles Duke's family snapshot still sits on the dust of this atmosphere-less Moon, although its image has, by all accounts, been bleached by solar radiation.

What these men were even less prepared for was the profound, transformative emotional impact of seeing the Earth from afar. And yet this features in the earliest narratives of lunar travel, in Lucian's *Ikaromenippos* or Johannes Kepler's *Somnium*, and views of Earth from the Moon recur in early moonscapes. Fritz Lang's *Frau im Mond* anticipated the emotion this sight elicited: the six passengers huddle around a porthole, gawping with awe, doe-eyed Friede sizzling with emotion. Even the villain temporarily loses himself in the wondrous sight.

When humankind first circumnavigated the Moon on Christmas Eve 1968, time briefly stood still as the Apollo 8 crew recited the opening of the book of *Genesis*: 'In the beginning God created the heaven and the earth ...'. Beyond abiding to a specific religious creed, they were, they suggested, witnessing the whole of Creation while their message was broadcast to the world. They also caught sight of Earth rising over the Moon's barren horizon. The Apollo astronauts had received strict instructions as to how they should spend their time on board, and how best to use the precious film in their cameras. Allowance had been made for 'targets of opportunity', which could be used by NASA's public affairs department, but were not a priority on the tight schedule of their mission. Yet, consciously or not, they knew that this had to be recorded, not just etched

○ *Apollo 16 astronaut Charles Duke's family snapshot on the Moon's surface, 20 April 1972*

on paper, circulated to the many. This had to be shared for others to see. This famous Earthrise was captured by Lunar Module Pilot William 'Bill' Anders, an image he has since described as 'paradigm shifting'.[3] It was published on the ront pages of newspapers across the world, and what an mpact it would have.

landscape of the Moon or from the dark void of space seeing the whole of humankind, past, present and future in that fragile orb protected by a thin layer of atmosphere elicited what space philosopher Frank White has called the 'overview effect'.[4] These men, as Collins puts it, were 'not trained to emote, [but] trained to repress emotions'.

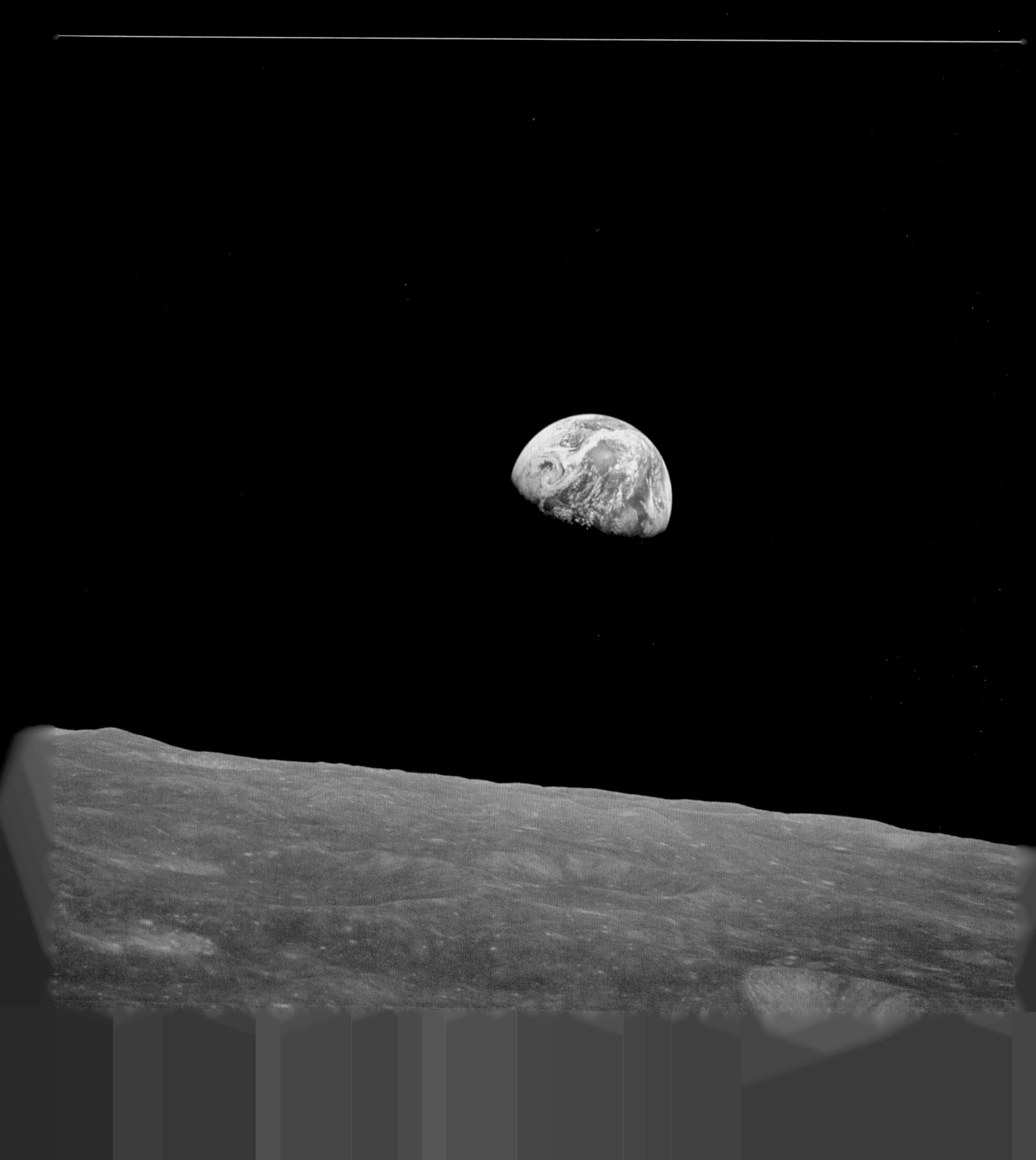

and yet their reaction to seeing the Earth from afar created a sentiment that oozes out of any interview, biography or Apollo narrative.[5] Perhaps because of the time of year at which he made his journey, Anders saw the Earth as a 'Christmas tree ornament'.[6] To his fellow astronauts, it was that 'beckoning oasis shining far away in the sky' (Aldrin), 'too beautiful to grasp' (Eugene Cernan), 'the garden of Eden' (Alan Bean), 'the most beautiful star in the heavens' (Cernan again), 'the only warm living object' in the immensity of space (James Irwin), 'the most beautiful, heart-catching sight of my life' (Frank Borman). Collins spoke of the 'tingle of awe' as he 'looked at it in wonderment, suddenly aware of how its uniqueness is stamped in every atom of my body'.[7] The Earth was us and we were the Earth. Going to the Moon not only revealed 'our place in the larger scheme of things' (Edgar Mitchell),[8] but also made 'you realize that on that small spot, that little blue and white thing, is everything that means anything to you – all of history and music and poetry and art and death and birth and love, tears, joys, games, all of it on that little spot out there that you can cover with your thumb' (Russell Schweickart).

These were troubled years, marked by the post-war divisions of the Cold War, a wave of decolonisation on the African continent and the cultural revolutions of 1968. The American Dream was shaken. Unease was growing over the Vietnam War, which grew bloodier, more senseless and lengthier by the day. Civil unrest was fuelled by poverty and racial discrimination. President John F. Kennedy and his brother, the would-be presidential candidate Robert F. Kennedy were assassinated. Martin Luther King Jr, leader of the Civil Rights movement, suffered the same fate. Gil Scott-Heron's spoken-word poem, 'Whitey on the Moon' (1970), expressed in no uncertain terms how the Apollo programme – its tremendous costs being borne by American citizens often living in squalor – was deflecting attention from shameful inequality.

Seeing the Earth from afar showed the incongruity of political boundaries. There was, they found, one Earth, one human species, and this pale blue marble was their home, our home, the only home. The astronauts saw themselves as apostles, there to deliver a message before it was too late. Collins poignantly believes that:

> *if the political leaders of the world could see their planet from a distance of 100,000 miles, their outlook could be fundamentally changed. That all-important border would be invisible, that noisy argument suddenly silenced. The tiny globe would continue to turn, serenely ignoring its subdivisions, presenting a unified face that would cry out for unified understanding, for homogeneous treatment.*[9]

At the end of the Apollo programme, when the NASA administrators reflected on its achievements, they recognised that it was towards the Earth that the mission had truly turned our sights. Apollo 8's *Earthrise* has been credited with kick-starting the environmental movement. Apollo 17's *Blue Marble*, a picture of the whole Earth in full illumination, became a symbol of environmental activism and of the annual Earth Day celebrations. In the years following the end of Apollo, NASA became key in monitoring environmental changes, with projects such as Landsat, and the discovery of the hole in the ozone layer over Antarctica.

One could feel disillusioned that, despite the initial hope and optimism that to look back at the Earth would help us treat our home better, 50 years on, this has, by all accounts, failed. Global carbon emissions are at an all-time high, temperatures keep rising, and with climate change, humankind is hurling itself into the worst crisis it has ever faced, a crisis that will most certainly bring about land and food shortages, mass migration and its likely corollaries, disease and conflict. Despite the recognition that the Anthropocene has brought about inexorable changes, the Kyoto and Paris Agreements to limit these damages have suffered setbacks. Capitalism, it transpires, has more clout than environmentalism. As we set our sights on the Moon again, whether to see it as a springboard for human exploration (and colonisation) of Mars or as a site of exploitation (with the ethical concerns that go with this), it is time, too, to look back at the Earth.

Upon returning home, Anders proclaimed: 'we came all this way to explore the Moon, and the most important thing is that we discovered the Earth'. The dream of going to another world turned out to be not about the journey or the destination, but the return. One of the questions this book asks the reader is, 'what is the Moon to you?' As we celebrate the 50th anniversary of humankind's first steps on another world, this is, more crucially, the time to pause and consider: what will you do for your Earth? As Michael Collins has said: 'Let us treasure it. There is not another one.'

Whole Earth Epilog

access to tools

U.S.A. $4
Canada $5
United Kingdom £1.75
Australia $4 (recommended)
New Zealand $4

First Edition
September 1974

○ *The Apollo 11 lunar module, Eagle, returning from the Moon to dock with the command module* Columbia, *21 July 1969*

Endnotes

A CONSTANT COMPANION

Using the Moon

1 John Conduitt, 'Notes on Newton's character', King's College, Cambridge, Keynes MS 130.07, fol. 6v, quoted in Richard Dunn and Rebekah Higgitt, *Finding Longitude* (Glasgow: Collins, 2014), p. 56.

2 Thomas Brisbane, speech at Glasgow Observatory, 16 December 1836, State Library of New South Wales, Mitchell MSS 1191/1/521, quoted in Dunn and Higgitt, *Finding Longitude*, p. 156.

3 John Worlidge, *Systema Agriculturae* (London: J.C. for T. Dring, 1675), p. 307.

4 William Shakespeare, *King Lear*, ed. by Jay L. Halio (Edinburgh: Oliver & Boyd, 1973), p. 29 (Act 1:2, lines 94–5).

5 William Ramesey, *Astrologia Restaurata or Astrology Restored* (London: Printed for Robert White, 1653), p. 63.

6 Simon Forman, 'The Astrologicalle Judgmentes of phisick and other Questions', in *The Casebooks Project* <http://www.magicandmedicine.hps.cam.ac.uk/view/text/normalised/TEXT5> [accessed 20 August 2018].

7 Marsilio Ficino, *Three Books on Life*, ed. and trans. by C. Kaske and J. Clark (Binghamton: MRTS, 1989), p. 316.

8 Nicholas Culpeper, *The English Physician Enlarged with Three Hundred, Sixty and Nine Medicines Made of English Herbs* (London: Peter Cole, 1653), pp. 51 and 58.

'Never Point at the Moon!': Lunar Lore across the Millennia

1 Richard H. Wilkinson, *The Complete Gods and Goddesses of Ancient Egypt* (London, 2003), p. 114.

2 Robert Ritner, 'Anubis and the Lunar Disk', *The Journal of Egyptian Archaeology*, 71 (1985), pp. 149–55 (p. 152).

3 Ritner, p. 151.

4 Richard A. Proctor, *Myths and Marvels of Astronomy* (London, 1878), pp. 246–47.

5 William Shakespeare, *A Midsummer Night's Dream*, ed. by Sakhunta Chaudhury (London: Bloomsbury, 2017), p. 266 (Act 5:1, lines 251–54).

6 Aristophanes, *The Clouds*, trans. by Alan Sommerstein (London: Penguin, 1973), pp. 143–44 (lines 749–56).

7 Plutarch, 'De defecto oraculorum', in *Plutarch's Morals*, revised by William W. Goodwin, 5 vols. (Boston: Little, Borown & Co., 1874), vol. 4, p. 18.

8 William Shakespeare, *The Tempest*, ed. by David Lindley (Cambridge: Cambridge University Press, 2002), p.214 (Act 5:1, lines 268-70).

9 J. Rotton and I.W. Kelly, 'Much ado about the full moon: A meta-analysis of lunar-lunacy research', *Psychological Bulletin*, 97:2 (1985), pp. 286–306.

10 Charles L. Raison, Haven M. Klein and Morgan Steckler, 'The Moon and Madness Reconsidered', *Journal of Affective Disorders*, 53:1 (1999), pp. 99–106. Summarised in S.O. Lilienfold and H. Arkovitz, 'Lunacy and the Full Moon', *Scientific American Mind*, 20:1 (2009), pp. 64–5, (p. 65).

11 Lady Jane 'Speranza' Wilde, *Ancient Legends, Mystic Charms, and Superstitions of Ireland* (Boston: Ticknor, 1887), p. 170.

12 Gervaise of Tilbury, Otia Imperialia (c. 1211), quoted in D.I. Shyovitz, 'Christians and Jews in the Twelfth-Century Werewolf Renaissance', *Journal of the History of Ideas*, 75:4 (2014), pp. 521–43, (p. 540).

13 Bram Stoker, *The New Annotated Dracula*, ed. by L. Klinger (London, 2008), p. 339.

The Moon in China

1 I use this term to refer to the succession of variously named separate states and empires on the East Asian landmass, nowadays seen as ancestral to the modern Chinese state.

Lunar Illumination in the Art of Africa

1 Anthony F. Aveni, *Conversing with the Planets: How Science and Myth Invented the Cosmos* (New York: Times Books, 1992), p. xi.

2 Drid Williams, 'The Dance of the Bedu Moon', in *African Arts*, 2:1 (1968), pp. 18–21 (p. 18).

3 Dunja Hersak, 'Colours, Stripes and Projection: Revelations on Fieldwork Findings and Museum

Enigmas', in Luc de Heusch (ed.), Objects, *Signs of Africa* (Ghent: Snoeck-Ducjau and Zoon), pp. 160–173. (1996) p. 163.

4 Mary Nooter Roberts and Allen F. Roberts, *Luba* (Milan: 5 Continents Editions, 2007), p. 47.

5 Allen F. Roberts, 'Social and Historical Contexts of Tabwa Art', in Evan M. Maurer and Allen F. Roberts (eds), *Tabwa: The Rising of a New Moon: A New Century of Tabwa Art* (Ann Arbor: University of Michigan Museum of Art, 1985), pp. 1–48 (p. 2).

6 Allen F. Roberts, 'Tabwa Masks: An Old Trick of the Human Race', in *African Arts*, 23:2 (1990), pp. 36–47, pp. 101–3, p. 39.

A Place that Exists Only in Moonlight: the Moon as Muse

1 The sketches are now in the Victoria and Albert Museum.

2 John Milton, *A Maske Presented at Ludlow Castle, 1634* ('Comus') (London: For Humphrey Robinson, 1637), p. 8, lines 219–20.

3 Samuel Palmer, *An English Version of the Eclogues of Virgil* (London: Seeley & Co., 1883), p. 82.

4 Sheena Wagstaff, 'Night and Day', in Darren Almond, *Fullmoon*, ed. by Hans Werner Holzwarth (Cologne: Taschen, 2014), pp. 7–9, (p. 9).

5 Darren Almond in conversation with William A. Ewing, *Photo London*, Somerset House, 17 May 2018.

6 Buzz Aldrin, in *Apollo 11, Technical Air-to-Ground Voice Transcription*, NASA, Manned Spacecraft Centre, Houston, Texas, July 1969, available online at https://www.hq.nasa.gov/alsj/a11/a11transcript_tec.pdf [accessed 23 September 2018].

7 Andrew Chaikin, in Michael Light, *Full Moon* (London: Jonathan Cape, 1999), unpaginated.

8 Alan Bean with Andrew Chaikin, *Apollo: An Eyewitness Account by Astronaut/Explorer Artist/Moonwalker Alan Bean* (Seymour CT: Greenwich Workshop, 1998), pp. 50–51.

The Moon and Music

1 Author's translation

2 This is a loose paraphrase from Alessandro Striggio's libretto:
Io la Musica son, ch'a i dolci accenti
so far tranquillo ogni turbato core,
ed or di nobil ira, ed or d'amore
posso infiammar le più gelate menti.

3 Füllest wieder Busch und Tal
Still mit Nebelglanz,
Lösest endlich auch einmal
Meine Seele ganz.
(Author's translation)

4 Timothy Jones, Beethoven: The *'Moonlight' and Other Sonatas, Op. 27 and Op. 31* (Cambridge: Cambridge University Press, 1999), p. 44

THROUGH THE LENS

The Telescopic Revolution

1 William Lower to Thomas Harriot, 6 February 1610, quoted in J. Bryn Jones, 'Lower, William', in Thomas Hockey et al. (eds), *Biographical Encyclopedia of Astronomers* (New York: Springer, 2007), p. 712.

2 Sir Henry Wotton to Robert Cecil, Earl of Salisbury, 13 March 1610, quoted in Eileen Reeves, *Painting the Heavens: Art and Science in the Age of Galileo* (Princeton: Princeton University Press, 1997), pp. 27–28.

3 Galileo Galilei, *Sidereus Nuncius* (Venice, 1610), quoted in Samuel Y. Edgerton, 'Galileo, Florentine "Disegno," and the "Strange Spottednesse" of the Moon', *Art Journal*, 44 (1984), pp. 225–32 (p. 229).

4 William Lower to Thomas Harriot, 11 June 1610, quoted in Scott L. Montgomery, *The Moon and the Western Imagination* (Tucson: University of Arizona Press, 1999), p. 112.

5 Martin Horky to Johannes Kepler, 27 April 1610, quoted in Albert Van Helden, 'Telescopes and Authority from Galileo to Cassini', *Osiris*, 9 (1994), pp. 8–29 (p. 11).

Mapping the Moon

1. Hugh P. Wilkins, *Mysteries of Space and Time* (London: Frederick Muller, 1955); p. 51, 52, 57.
2. Mary Winkler and Albert Van Helden, 'Representing the Heavens: Galileo and his Visual Astronomy', *Isis* 83 (1992), pp. 195–217.
3. Nydia Pineda de Avila 'Copying Hevelius's lunar template', *Making Visible Project Blog*, 7 September 2016 <http://www.mv.crassh.cam.ac.uk/2016/09/07/copying-heveliuss-lunar-template/> [accessed 10 May 2018].
4. Jean-Dominique Cassini, quoted in Luisa Pigatto and Valeria Zanini, 'Lunar Maps of the seventeenth and eighteenth centuries. Tobias Mayer's Map and its nineteenth-century edition,' *Earth, Moon and Planets*, 85–86 (1999), 365–377 (p. 377).
5. John Phillips, 'Suggestions for the Attainment of a Systematic Representation of the Physical Aspect of the Moon', *Philosophical Transactions of the Royal Society of London* 12 (1862–63), pp. 31–7 (p. 35).
6. William Pickering, *Investigations in Astronomical Photography* (Cambridge, MA: Annals of the Astronomical Observatory of Harvard College 32, 1895), p. 109.
7. *Edinburgh Review* (July 1874), pp. 74–5.
8. Norman Lockyer, quoted in Boris Jardine, 'Made real: artifice and accuracy in nineteenth-century scientific illustration', *Science Museum Group Journal* 2 (2014), pp. 1–32 (p. 17).
9. James Nasmyth and James Carpenter, *The Moon: Considered as a World, a Planet, and a Satellite* (London: John Murray, 1885), p. 202.
10. Nasmyth and Carpenter, p. 174.
11. Nasmyth and Carpenter, p. 178.
12. Adrien Auzout, 'Monsieur Auzout's Speculations of the Changes, Likely to be discovered in the Earth and Moon, by their respective inhabitants', *Philosophical Transactions of the Royal Society of London*, 1 (1665–6), pp. 120–3.
13. William Herschel, 'An account of three volcanoes on the Moon', *Philosophical Transactions of the Royal Society of London*, 77 (1787), pp. 229–32.
14. William Herschel to Nevil Maskelyne, 12 June 1780 in *The Scientific Papers of William Herschel*, ed. by J. Dreyer, 3 vols (Cambridge: Cambridge University Press, 1912), I, pp. xc–xci.
15. Quoted in William Sheehan, *Planets and Perception: Telescopic Views and Interpretations, 1609–1909*, (Tucson: University of Arizona Press, 1989), p. 33.
16. John P Nichol, *Contemplations on the Solar System* (Edinburgh: Tait, 1844), p. 154.
17. Nichol, p. 165.
18. Ewen Whittaker, *Mapping and Naming the Moon: A History of Lunar Cartography and Nomenclature* (Cambridge: Cambridge University Press, 2003), p. 55.
19. Will van den Hoonard, 'Moonstruck: Cartographic Explorations of the Moon by Mary A Blagg and Kira B Shingareva,' *Terrae Incognitae*, 48 (2016), pp. 76–86 (pp. 78–9).
20. Hugh Percival Wilkins, 'Notes on Lunar Drawing', *The Moon* 3:2 (1954), p. 43.
21. William Pickering, quoted in William Leatherbarrow, 'Hugh Percival Wilkins, 1896-1960: an appreciation' *Journal of the British Astronomical Association*, 120:1 (2010), pp. 39–42 (pp. 40–1).
22. William Pickering, quoted in Leatherbarrow, pp. 40–1.
23. 'Moon pictures show "monotonous" other side' *Science*, 130:3384 (1959), 1241–2.
24. Elly Dekker, *Globes at Greenwich: A catalogue of the globes and armillary spheres at the National Maritime Museum* (Oxford: Oxford University Press 1999), pp. 307–8.

Portraying the Moon: an Artist's Eye

1. Neil Jeffares, 'Russell, John: Part I, Essay and Sitters A-E' and 'Russell, John: Part VI, Unidentified Sitters &c.', *Dictionary of Pastellists before 1800, online edition*, www.pastellists.com [accessed 5 April 2018].
2. The sketches are at the Museum of the History of Science, Oxford. The pastels are at the National Maritime Museum, the Royal Astronomical Society and the Science Museum in London, Birmingham Museum and Art Gallery, and the Museum of the History of Science, Oxford. He also produced a 2-feet-diameter relief globe and relief segments for the *Selenographia*: W. F. Ryan, 'John Russell, R.A. and early lunar mapping', *The Smithsonian Journal of History*, 1 (1966), 27–48 (pp. 28, 39).
3. Oxford, Museum of the History of Science, MS Radcliffe 72/2: Professor Stephen P. Rigaud, 16 December 1824, *Notes on Russell's Drawing of the Moon, Notes by Professor Rigaud copied from the M.S. obligingly lent by his son Major General Rigaud by R. Main*, 1824–5, pp. 2–3.

4 John Russell quoted in Edward James Stone, 'Note on a Crayon Drawing of the Moon by John Russell RA at the Radcliffe Observatory, Oxford', *Monthly Notices of the Royal Astronomical Society*, 56 (1895), 88–95 (p. 91).

5 Russell in Stone, p. 94.

6 Russell in Stone, p. 92.

7 Russell in Stone, p. 91.

8 Joseph Farington, *The Diary of Joseph Farington*, Kenneth Garlick, Angus Macintyre and Kathryn Cave (eds), 16 vols (New Haven, London: Yale University Press, 1978–84), I, p. 110 (9 December 1793); VIII, p. 2887 (15 October 1806).

9 Russell in Stone, p. 92.

10 W. F. Ryan, 'John Russell, R.A. and early lunar mapping', *The Smithsonian Journal of History*, 1 (1966), 27–48 (p. 35).

11 Russell in Stone, p. 94.

12 Russell in Stone, pp. 91, 92.

13 John Russell, *Elements of Painting with Crayons*, (London: Wilkie, 1772), p. ii.

14 Russell in Stone, pp. 93–4.

15 Russell in Stone, p. 94.

16 W. F. Ryan, 'John Russell, R.A. and early lunar mapping', *The Smithsonian Journal of History*, 1 (1966), 27–48 (p. 45).

17 Russell in Stone, p. 92.

18 William Russell, *A Description of the Lunar Planispheres, engraved by the late John Russell from his original Drawings* (London: W. Russell, 1809), p. 2.

19 William Russell, p. 3.

20 Russell, *Elements*, p. 19.

21 Farington, III, p. 695 (11 November 1796).

22 Antje Matthews, 'John Russell (1746–1806) and the Impact of Evangelicalism and Natural Theology on Artistic Practice' (unpublished doctoral thesis, University of Leicester, 2005), pp. 147, 140.

23 Anon., "Remarks on the exhibition of paintings &c... at the Royal Academy", *St James's Chronicle*, May 1795, pp. 14–6, quoted in Jeffares, 'Russell, John: Part I', p. 4.

24 Russell in Stone, p. 95.

25 *Selenographia* on simple mount, RAS524. Epigram in Latin, 'In Johanem Russelium Lunae pictorem / Ne propre viderrunt Actaeon Endymionque / Hos memini solos; ast ubi Russelius ?'; translation by Ryan, p. 44.

By the Light of the Moon

1 Alfred Brothers, 'Astronomical Photography', *British Journal of Photography*, 30:1200 (1883), 248–249 (p. 248).

2 'Fixation des images qui se forment au foyer d'une chambre obscure', *Comptes Rendus* 8:1 (7 January 1839), pp. 4–7.

3 Theresa Levitt, 'Biot's Paper and Arago's Plates: Photographic Practice and the Transparency of Representation', *Isis* 94:3 (September 2003), pp. 456–76 (pp. 456–57).

4 Michael Robinson, *The Techniques and Material Aesthetics of the Daguerreotype* (Leicester: De Montfort University, 2017), pp. 74–93, <https://www.dora.dmu.ac.uk/xmlui/handle/2086/14332> [accessed 14 June 2018].

5 JFW Herschel to WHF Talbot, 30 January 1839. *The Correspondence of William Henry Fox Talbot*, Letter number 03780 <http:foxtalbot.dmu.ac.uk> [accessed 10 June 2018].

6 WHF Talbot to JFW Herschel, 15 November 1839. The Correspondence of *William Henry Fox Talbot*, Letter number 03971 <http:foxtalbot.dmu.ac.uk> [accessed 10 June 2018].

7 JFW Herschel to WHF Talbot, 4 December 1839, *The Correspondence of William Henry Fox Talbot*, Letter number 03982 <http:foxtalbot.dmu.ac.uk> [accessed 10 June 2018].

8 WHF Talbot to JFW Herschel, 1 July 1841. *The Correspondence of William Henry Fox Talbot*, Letter number 04293 <http:foxtalbot.dmu.ac.uk> [accessed 10 June 2018].

9 Michael Robinson, *The Techniques and Material Aesthetics of the Daguerreotype* (Leicester: De Montfort University, 2017), pp. 216–25. <https://www.dora.dmu.ac.uk/xmlui/handle/2086/14332> [accessed 14 June 2018].

10 Dorrit Hoffleit, *Some Firsts in Astronomical Photography* (Cambridge, Massachusetts: Harvard College Observatory, 1950), pp. 25–26.

11 Omar Nasim, 'James Nasmyth on the Moon; Or on Becoming a Lunar Being Without the Lunacy', in *Selene's Two Faces: From 17th Century Drawings to Spacecraft Imaging*, ed. by Carmen Pérez González (Leiden: Brill, 2018), pp. 147–187 (p. 151).

12 *Official descriptive and illustrated catalogue of the Great exhibition of the works of industry of all nations,*

1851. (London: Spicer Brothers, 1851). Exhibit 296.16(6), v1, Class 10, p. 441.

13 Known daguerreotypes of the moon were attempted by Samuel D. Humphreys, Whipple and Bond, Daguerre, Draper, Claudet, and the author of an anonymous stereo daguerreotype of the moon in a collection in Hamburg, Germany. [www.daguerreobase.org L196].

14 M. L. Huggins, 'Obituary: Warren De La Rue', *The Observatory*, 12 (1889), pp. 245–250.

15 Patent numbers COPY 1/4/435; 1/1/163; 1/4/436; 1/1/203.

16 See for instance the entry for the George Eastman Museum's inventory number 1981.8775.0017 <https://collections.eastman.org> [accessed 20 July 2018].

17 For a full account see Omar Nasim, pp. 160–161.

18 For technical details of the different processes, see John Towler's handbook, *The Silver Sunbeam* (New York: J.H. Ladd, 1864).

19 For more about the observatory see J. L. Birks, 'The Penllergare Observatory', *Antiquarian Astronomer*, 2005, Issue 2, pp. 3–8.

20 Richard L. Maddox, 'An Experiment with Gelatino-Bromide,' *The British Journal of Photography*, 18:592 (1871), 422–423.

21 *Exhibitions of the Royal Photographic Society 1870-1915*. <http://erps.dmu.ac.uk>. [accessed 10 June 2018].

22 Monique Sicard, 'L'Atlas photographique de la Lune, de MM. Loewy et Puiseux', *Revue de la BNF*, 44:2 (2013), pp. 36–43 <https://www.cairn.info/revue-de-la-bibliotheque-nationale-de-france-2013-2-page-36.htm> [accessed 10 August 2018].

23 Sicard, p. 8.

24 Lorraine Daston and Peter Galison, *Objectivity* (London: Zone Books, 2006).

The Moon through the Eyes of the Apollo Astronauts

1 Eugene Cernan, interviewed by Rebecca Wright, December 11, 2007. Interview Transcript, NASA Johnson Space Center Oral History Project, Houston, TX, p. 32.

2 *New York Times*, 24 December, 1972, front page; *Boston Globe*, 25 December, 1972, front page.

3 Alison Landsberg, *Prosthetic Memory: The Transformation of American Remembrance in the Age of Mass Culture* (New York: Columbia University Press, 2004).

DESTINATION MOON

The End of the Beginning: Imagining the Lunar Voyage

1 Publicity brochure for *Destination Moon*, quoted in David Seed, 'Atomic Culture and the Space Race', in *The Oxford Handbook of Science Fiction*, ed. by Rob Latham (Oxford: Oxford University Press, 2014), pp. 340–51 (p. 343).

Project Moon: Satirising our Satellite

1 Ludovico Ariosto, *Orlando Furioso*, trans. by John Hoole, 5 vols. (2nd edition; London: Printed for George Nicol, 1785), vol. 4, pp. 217–18.

2 John Carswell, *The South Sea Bubble* (London: Cresset Press, 1960).

3 David Dabydeen, *Hogarth, Walpole and Commercial Britain* (London: Habsib, 1987), p. 51.

4 *Morning Post and Daily Advertiser*, 6 February 1784.

5 My thanks to Clare Brant for her advice on this print. See Clare Brant, *Balloon Madness: Flights of Imagination in Britain, 1783–1786* (Woodbridge: Boydell Press, 2017).

6 James Taylor, *Creating Capitalism: Joint-Stock Enterprise in British Politics and Culture 1800–1870* (London: Royal Historical Society, 2006), pp.93–7. My thanks to Anton Howes for his thoughts on this print.

7 Samuel Tipper, *The Satirist, or Monthly Meteor*, 1 October 1809, pp. 316–17.

8 Frank M. Parisi, 'Emblems of Melancholy. For Children: The Gates of Paradise', in Michael Phillips (ed.), *Interpreting Blake* (Cambridge: Cambridge University Press, 1978), pp. 86–7.

Life on the Moon, Newspapers on Earth

1 'Great Astronomical Discoveries', *Sun*, 26–31 August 1835.

2 'Richard Adams Locke' (1846), in Edgar Allan Poe, *Essays and Reviews*, ed. by G. R. Thompson (New York: Literary Classics of the United States, 1984), p. 1220.

3 For a critical account, Michael J. Crowe, 'William and John Herschel's Quest for Intelligent Life', in Clifford J. Cunningham (ed.), *The Scientific Legacy of William Herschel* (Cham, Switzerland: Springer, 2018), pp. 239–74 (pp. 261–5).

4 Richard Adams Locke, *New World*, 16 May 1840, p. 1.

5 'Discoveries in the Moon', *Mercantile Advertiser*, 28 August 1835, p. 2.

[6] 'Great Astronomical Discoveries', *New-York Daily Advertiser*, 28 August 1835, p. 2.

[7] 'Great Astronomical Discoveries', *Evening Post*, 28 August 1835, pp. 1–2.

[8] 'Editorial', *Evening Post*, 29 August 1835, p. 1.

[9] 'Stupendous Discovery in Astronomy', *Albany Daily Advertiser*, quoted in *Sun*, 1 September 1835, p. 2.

[10] 'Ass-tronomer Herschel to the Satirist', *Satirist*, 24 April 1836, p. 131.

Fly Me to the Moon: From Artistic Moonscape to Destination

[1] William Herschel, 'Moon', 28 May 1776, Royal Astronomical Society, London, RAS MSS Herschel W. 3/1.4. *Moon* (7 March 1775–9 November 1807), p. 2.

[2] Josiah Crampton, *The Lunar World: Its Scenery, Motions Etc. (Considering with a View to Design)* (Dublin: George Herbert, 1853), p. 21.

[3] Crampton, p. 21.

[4] James Nasmyth and James Carpenter, *The Moon: Considered as a Planet, a World, a Satellite* (London: John Murray, 1874), pp. 162–66.

[5] Camille Flammarion, *Les Terres du Ciel* (Paris: Didier et Cie., 1877), plates facing p. 536 and p. 481.

[6] Flammarion, plate facing p. 532.

[7] Flammarion, plate facing p. 36.

[8] Ron Miller, 'Chesley Bonestell, the Fine Art of Space Travel', in *A Chesley Bonestell Space Art Chronology*, ed. by Melvin H. Schuetz (Parkland, FL: Universal Publishers, 1999), p. xix.

[9] Quoted in Fredrick C. Durant III and Ron Miller (eds.), *Worlds Beyond: The Art of Chesley Bonestell* (Norfold, VA: Donning, 1983), p. 8.

[10] Robert A. Heinlein, 'Shooting "Destination Moon"' in *Astounding Science Fiction*, July 1950, quoted in Durant and Miller, p. 11.

[11] 'Destination Moon (1950); The screen: Two new features arrive: "Destination Moon," George Pal version of "Rocket Voyage"', *The New York Times*, 28 June 1950.

[12] Miller, p. xxi.

[13] Miller, p. xxiii.

[14] Miller, p. xxv.

[15] 'Vija Celmins – Artist Rooms | TateShots', https://www.youtube.com/watch?v=SsbkzSrCdlg [accessed 12 June 2018].

[16] W.H. Auden, 'Moon Landing', in *Selected Poems* (Expanded 2nd edition, New York: Vintage Books, 2007), p. 307.

[17] Kiki Kogelnik, *Kiki Kogelnik: Fly me to the Moon*, ed. by Ciara Moloney (Oxford: Modern Art Oxford, 2015), p. 45.

[18] Quoted in Alexandra Hennig, 'Robot Fantasies, or How the Bombs Learned to Love', in Hans Peter Wipplinger (ed.), *Kiki Kogelnik: Retrospektive/ Restrospective* (Krems: Kunsthalle Krems, 2013), pp. 34–45 (p. 39).

[19] Robert Rauschenberg, 'Notes on Stoned Moon', *Studio International*, 178:197 (December 1969), pp. 246–47 (p. 247).

[20] Bryony Fer, in Achim Borchardt-Hume (ed.), *Rothko: The Late Series* (London: Tate Publishing, 2008), pp. 31–43 (p. 33).

[21] Fer, p. 33.

Fashion and the Moon Landings

[1] Kathleen Halton, 'Tom Stoppard: The startling young author of the play *Rosencrantz and Guildenstern are Dead*', [American] Vogue, 15 October 1967, pp. 112–13.

[2] Nicholas de Monchaux, *Spacesuit: Fashioning Apollo* (Cambridge, USA: MIT Press, 2011), p. 325.

[3] De Monchaux, p. 127.

[4] Revlon Inc. advertisement, 'Moon Drops', [American] *Vogue*, 1 September 1969, pp. 108–9.

FOR ALL MANKIND?

So What If It's Just Green Cheese? The Moon on Screen

[1] Cynthia Lowry, 'TV Fans Save Space Ship Enterprise From Mothballs', *Florence Times*, 29 March 1968, p. 15.

[2] Michael Billington, 'Moonmen or Dr. Finlay', *The Times*, 21 July 1969, p. 11.

[3] Oliver Postgate, *Seeing Things: A Memoir* (Edinburgh: Canongate Books, 2009), p. 279.

Poetry and the Moon: Moon Vehicles, Moon Metaphors

[1] Sir Philip Sidney, *The Major Works* (Oxford: Oxford University Press, 2008), p. 165.

[2] Quoted in Henry Little, 'For All Mankind: A Brief Cultural History of The Moon', in *The White Review*, September 2013 <www.thewhitereview.org/feature/

for-all-mankind-a-brief-cultural-history-of-the-moon> [accessed 24 August 2018].

3 Quoted in Little, 2013.

4 John Milton, *Paradise Lost* (New York: W.N. Norton & Company, 1993/1674), p. 84.

5 Liane Strauss, *All the Ways in Which You Still Remind Me of the Moon* (London: Paekakariki Press, 2015), p. 6.

6 Rebecca Elson, *A Responsibility to Awe* (Manchester: Carcanet, 2001), p. 23.

7 There are countervailing arguments that speak of an equilibrium eventually being reached to stay this process.

Whose Moon? The Artists and the Moon

1 Quoted in Leslie Katz, 'Scratch and Sniff this Art for a Waft of the Moon', 22 October 2010, in *cnet.com*, <https://www.cnet.com/news/scratch-and-sniff-this-art-for-a-waft-of-moon/> [accessed 30 August 2018].

2 Kenneth Zin, Eng. Electronic, Lunar Orbiter Image Recovery, Mac Moon, NASA Ames Research Center, speaking in Nelly Ben Hayoun, 'Moon Dust Remix by Nelly Ben Hayoun_the Sound of Neil Armstrong Boots on the Moon', in *Vimeo*, <https://vimeo.com/53524897> [accessed 30 August 2018].

3 Nelly Ben Hayoun, 'Moon Dust Remix Soundtrack', in *Vimeo*, <https://vimeo.com/56696945> [accessed 30 August 2018].

4 Aleksandra Mir, 'First Woman on the Moon', in *Aleksandra Mir*, <https://www.aleksandramir.info/projects/first-woman-on-the-moon/> [accessed 30 August 2018].

5 Alicia Framis, 'Moon Academy', in *Alicia Framis*, <http://aliciaframis.com.mialias.net/2010-2/moon-academy-aprilaugust-2010-experimental-academy-in-amsterdam/> [accessed 30 August 2018].

6 'Artist Mikael Genberg to Install Self-assembling House on the Moon in 2015', in *designboom*, <https://www.designboom.com/technology/moonhouse-project-mikael-genberg-to-install-self-assembling-house-on-the-moon-05-29-2014/> [accessed 30 August 2018].

7 'Republic of the Moon', in *The Arts Catalyst*, <https://artscatalyst.org/republic-moon-london> [accessed 30 August 2018].

The Geopolitics of the Moon

1 United Nations, 'Treaty on Principles Governing the Activities of States in the Exploration and Use of Outer Space, including the Moon and Other Celestial Bodies', RES 2222 (XXI), 1966 <http://www.unoosa.org/oosa/en/ourwork/spacelaw/treaties/outerspacetreaty.html> [accessed 23 September 2018].

2 Robert Goddard, *A Method of Reaching Extreme Altitudes* (Washington DC: Smithsonian Institution, 1919), p. 57.

3 Quoted in William E. Burrows, *This New Ocean: The Story of the First Space Age* (New York: The Modern Library, 1998), p. 46.

4 'Topics of the Times', *New York Times*, 13 January 1920, p. 12.

5 'Soviet Fires Earth Satellite into Space; It is Circling the Globe at 18,000 M.P.H.; Sphere Tracked in 4 Crossings Over U.S.', *The New York Times*, 5 October 1957.

6 https://www.nasa.gov/multimedia/podcasting/jpl-sputnik-20071002.html [accessed 23 September 2018]

EPILOGUE

We Came All This Way

1 Quoted in Robert Poole, *Earthrise: How Man First Saw the Earth* (New Haven and London: Yale University Press, 2008), p. 106.

2 NASA, 'Apollo 11 – Air-to-ground voice transcription, July 16–24 1969 (*Goss Net 1*)', Manned Spacecraft Centre, Houston, July 1969, Tape 7-/26, p. 379.

3 William 'Bill' Anders, interview with the author, August 2015.

4 Frank White, *The Overview Effect: Space Exploration and Human Evolution* (Reston VA: American Institute of Aeronautics & Astronautics, 1998).

5 Unless otherwise attributed, this and subsequent quotes from astronauts are from Poole's *Earthrise*, as is much of the argument about the impact of seeing the Earth from afar.

6 Francis French and Colin Burgess, *In the Shadow of the Moon: A Challenging Journey to Tranquility, 1965–1969* (Lincoln NE: University of Nebraska Press, 2007), p. 308.

7 Michael Collins, 'Our planet: fragile gem in the universe', *Birmingham Post Herald*, 1 March 1972.

8 Edgar Mitchell, 1996, quoted in Piers Bizony, *Moonshots: 50 Years of NASA Space Exploration seen through Hasselblad Cameras* (Minneapolis: Voyageur Press, 2017), p. 141.

9 Michael Collins, *Carrying The Fire: An Astronaut's Journey* (New York: Cooper Square Press, 2001), p. 470.

○ *'Moon' square scarf in off-white silk twill, by Dior*

Further Reading

Al-Rodhan, Nayef R. F., *Meta-Geopolitics of Space: An Analysis of Space Power, Security, and Governance* (Basingstoke: Palgrave Macmillan, 2012).

Attlee, James, *Nocturne: A Journey in Search of Moonlight* (London: Penguin Books, 2012).

Aveni, Anthony F., *Ancient Astronomers* (Montreal and Washington D.C.: St. Remy Press and Smithsonian Books, 1993).

Benson, Michael, *Cosmigraphics: Picturing Space Through Time* (New York: Abrams, 2014).

Bizony, Piers, *Moonshots: 50 Years of NASA Space Exploration Seen Through Hasselblad Cameras* (Voyageur Press, 2017).

Brunner, Bernd, *Moon: A Brief History* (New Haven: Yale University Press, 2010).

Campbell, Dallas, *Ad Astra: An Illustrated Guide to Leaving the Planet* (London: Simon & Schuster, 2017).

Cashford, Jules, *The Moon: Myth and Image* (London: Cassell Illustrated, 2003).

Copenhaver, Brian P., *Magic in Western Culture: From Antiquity to the Enlightenment* (Cambridge: Cambridge University Press, 2015).

Crowe, Michael, *The Extraterrestrial Life Debate, 1750–1900* (Cambridge: Cambridge University Press, 1986).

Cullen, Christopher, *Heavenly Numbers: Astronomy and Authority in Early Imperial China* (Oxford: Oxford University Press, 2017).

Curry, Patrick, *Prophecy and Power: Astrology in Early Modern England* (Cambridge: Polity Press, 1989).

Dean, James D. and Ulrich, Bertram, *NASA/Art: 50 Years of Exploration* (New York: HNA Books, 2008).

Duffy, Carol Ann (ed.), *To the Moon: An Anthology of Lunar Poems* (London: Picador, 2009).

Duncan, David Ewing, *The Calendar* (London: Fourth Estate, 1998).

Dunn, Richard and Higgitt, Rebekah, *Finding Longitude* (Glasgow: HarperCollins, 2014).

Dunn, Richard, *The Telescope: A Short History* (London: National Maritime Museum, 2009).

Durant III, Fredrick C. and Miller, Ron (eds), *Worlds Beyond: The Art of Chesley Bonestell* (Norfolk, VA: Donning, 1983).

Freyer Stowasser, Barbara, *The Day Begins at Sunset: Perceptions of Time in the Islamic World* (London and New York: I.B.Tauris, 2014).

James, Jamie, *Music of the Spheres: Music, Science, and the Natural Order of the Universe* (Copernicus, 1993).

Kreamer, Christine Mullen (ed.), *African Cosmos: Stellar Arts* (Washington D.C. and New York: National Museum of African Art, Smithsonian Institution in association with The Monacelli Press, 2012).

Leatherbarrow, Bill, *The Moon* (London: Reaktion Books, 2018).

Light, Michael, *Full Moon* (London: Jonathan Cape, 1999).

Loske, Alexandra, and Massey, Robert, *Moon: Art, Science, Culture* (London: Ilex Press, 2018).
Mailer, Norman, *Moonfire, the Epic Journey of Apollo 11* (Cologne: Taschen, 2015).
Malin, David, and Murdin, Paul (ed.), *Universe: Exploring the Astronomical World* (London: Phaidon, 2017).
McEvoy, J.P., *Eclipse: The Science and History of Nature's Most Spectacular Phenomenon* (London: Harper Collins, 1999).
Miller, Ron, *The Art of Space* (Zenith Press, 2014).
Monchaux, Nicholas de, *Spacesuit: Fashioning Apollo* (Cambridge, USA: MIT Press, 2011).
Montgomery, Scott L., *The Moon and the Western Imagination* (Tucson: University of Arizona Press, 1999).
Milligan, Tony and Newman, Kim, *Nobody Owns the Moon: The Ethics of Space Exploitation* (Jefferson, North Carolina: McFarland & Co., 2015).
O'Brien, Daniel, *SF:UK How British Science Fiction Changed the World* (London: Reynolds and Hearn Ltd., 2000).
Parker Pearson, Mike, *Stonehenge: Exploring the Greatest Stone Age Mystery* (London: Simon & Schuster, 2013).
Pérez González, Carmen (ed.), *Selene's Two Faces: From 17th Century Drawings to Spacecraft Imaging* (Leiden: BRILL, 2018).
Poole, Robert, *Earthrise: How Man First Saw the Earth* (New Haven: Yale University Press, 2008).
Reynolds, David West, *Apollo: The Epic Journey to the Moon, 1963–1972* (Minneapolis: Zenieth, 2013).
Rothery, David A., *Moons: A Very Short Introduction* (Oxford: Oxford University Press, 2015).
Rydal Jørgensen, Laerke, and Laurberg, Marie (eds), *The Moon: From Inner Worlds to Outer Space* (Copenhagen: Louisiana Museum of Modern Art, 2018).
Stroud, Rick, *The Book of the Moon* (London: Doubleday, 2009).
Sheehan, Michael, *The International Politics of Space: No Final Frontier* (Abingdon: Routledge, 2007).
Smith, Andrew, *Moondust: In Search of the Men Who Fell to Earth* (London: Bloomsbury Publishing, 2009).
Spudis, Paul, *The Value of the Moon: How to Explore, Live, and Prosper in Space Using the Moon's Resources* (Washington DC: Smithsonian Books, 2016).
Whitehouse, David, *The Moon: A Biography* (London: Weidenfeld & Nicholson, 2016).
Whittaker, Ewen A., *Mapping and Naming the Moon: A History of Lunar Cartography and Nomenclature* (Cambridge: Cambridge University Press, 1999).
Williams, Edgar, *Moon: Nature and Culture* (London: Reaktion Books, 2014).
Young, Mark S., Duin, Steve, and Richardson, Mike, *Blast Off! Rockets, Robots, Ray Guns and Rarities from the Golden Age of Space Toys* (Milwaukie: Dark Horse, 2001).

Index

Note: page numbers in **bold** refer to information contained in captions, page numbers in *italics* refer to information contained in tables.

Authors

Megan Barford
Curator of Cartography at Royal Museums Greenwich. Her work is concerned with maps as cultural artefacts, and the way maps figure in, and figure, a variety of human relationships. This includes investigation into the range of different objects understood as maps, and the variety of practices that go into their making and use.

Simon Barraclough
Simon Barraclough's poetry collections include *Los Alamos Mon Amour* (Salt 2008), *Bonjour Tetris* (Penned in the Margins 2010), *Neptune Blue* (Salt 2011) and a book-length meditation on the sun, entitled *Sunspots* (Penned ... 2015). In 2014 Simon was writer in residence at the Mullard Space Science Laboratory. Simon is working on a collection of stories set in London.

Katy Barrett
Katy is currently Curator of Art Collections at the Science Museum, London. Katy was previously Curator of Art (pre-1800), at Royal Museums Greenwich, and has held various posts working across art and science collections. She has recently co-authored *The Sun: One Thousand Years of Scientific Imagery* (2018).

Scott Burnham
Distinguished Professor of Music, Graduate Center, City University of New York and Scheide Professor of Music History Emeritus, Princeton University, Scott began his musical life playing keyboard in a rock band but is better known today as the author of books and essays about Beethoven, Mozart, and Schubert.

Susanna Cordner
Susanna is a fashion curator and historian with a particular interest in the ways in which clothes can be used to explore personal and public experience. She runs the Archives at the London College of Fashion, University of the Arts London. She previously worked at the Victoria and Albert Museum, where she contributed to exhibitions such as *Undressed: A Brief History of Underwear*.

Christopher Cullen
Emeritus Director, Needham Research Institute, and Emeritus Fellow of Darwin College, Cambridge. Christopher has a long series of publications on the history of astronomy in China. His most recent is *Heavenly Numbers: Astronomy and Authority in Early Imperial China* (2017).

Louise Devoy
Louise is currently Senior Curator of the Royal Observatory at Royal Museums Greenwich. Louise has previously worked at the National Space Centre, Science Museum, British Museum and the National Air and Space Museum, Smithsonian Institution. She has visited many historic observatories worldwide and is fascinated by the people, places, objects and stories associated with the history of astronomy.

Richard Dunn
Richard worked at Royal Museums Greenwich between 2004 and 2019, mainly on the history of navigation and related subjects. His publications include *The Telescope: A Short History* (2009), *Finding Longitude* (2014, with Rebekah Higgitt) and *Navigational Instruments* (2016). He is currently Keeper of Technologies and Engineering at the Science Museum, London.

Simon Guerrier
A writer and producer, Simon studied GCSE astronomy as a night class at the Royal Observatory and quickly applied what he learned to the Doctor Who books and audio plays he writes. His books include *Top Trumps: Space* and *Doctor Who: The Women Who Lived*, and his most recent documentary, 'Victorian Queens of Ancient Egypt,' was broadcast on Radio 3.

John J. Johnston
John is a freelance Egyptologist, Classicist, and cultural historian. A former Vice-Chair of the Egypt Exploration Society, he has lectured extensively at many major institutions throughout the UK. His research interests include mortuary belief and practice, Hellenistic and Roman Egypt, the history of Egyptology, and the reception of ancient Egypt

in the modern world. In addition to contributing numerous articles to both academic and general publications, he has co-edited the following books: *Narratives of Egypt and the Ancient Near East: Literary Linguistic Approaches* (2011), *A Good Scribe and an Exceedingly Wise Man* (2014), and a major anthology of classic mummy fiction, *Unearthed* (2013).

Tom Kerss
Tom is a professional astronomer at the Royal Observatory Greenwich. Author of *Moongazing: Beginner's Guide to Exploring the Moon* (2018), he is one of the Royal Observatory's lead experts on the Moon. A keen stargazer and astrophotographer, he spends his days (and nights) communicating astronomy to students and the public and chasing clear skies.

Christine Mullen Kreamer
Christine is deputy director and chief curator at the National Museum of African Art, Smithsonian Institution. A life-long star-gazer, she is author and curator of *African Cosmos: Stellar Arts* (2012) and a host of other publications devoted to the classical and contemporary arts of Africa.

Jennifer Levasseur
Jennifer serves as Curator of Astronaut Personal Equipment and Cameras at the Smithsonian National Air and Space Museum. Her research includes an upcoming book on the cultural legacy of astronaut photography. She is also curator for the Museum's exhibition on post-Apollo human spaceflight, *Moving Beyond Earth*.

Will Newton
Currently a curator at the Victoria and Albert Museum of Childhood, the Moon, outer space and science fiction are among Will's favourite topics, so he delights in his responsibility for the Museum of Childhood's large collection of space toys.

David A. Rothery
Professor of Planetary Geosciences at the Open University, David chairs the Open University's level 2 module on planetary science and is the lead educator on free online learning packages about moons (our Moon and other moons!) offered by both FutureLearn and the Open University. He is on the team sending ESA's BepiColombo mission to Mercury.

James Secord
Professor of History and Philosophy of Science and director of the Darwin Correspondence Project at the University of Cambridge, his work on the history of science in the eighteenth and nineteenth centuries has been wildly published. His most recent book is *Visions of Science: Books and Readers at the Dawn of the Victorian Age* (2014).

Jill Stuart
Jill is an expert in the politics, ethics and law of outer space exploration and exploitation, based at the London School of Economics and Political Science. Within her area of research, the Moon has always provided a fascinating case study. From 2013–17 she was Editor in Chief of the Elsevier journal *Space Policy* where she remains a member of the Editorial Board.

Melanie Vandenbrouck
Curator of Art (post-1800) at Royal Museums Greenwich, Melanie is the lead curator of the exhibition *The Moon* (19 July 2019 – 5 January 2020), and is particularly interested in the collision between art and science. A long-time lover of the Moon, she has found a new fascination for astronomy thanks to the Insight Investment Astronomy Photographer of the Year competition of which she is a judge.

Kelley Wilder
Kelley is Director of the Photographic History Research Centre and Interim Director of the Institute of Art and Design at De Montfort University. She has research interests in the cultures of science and knowledge generated by photography and photographic practice. New projects include work on photographic catalogues and archives, and nineteenth and twentieth century material cultures of photographic science and industry.

Credits

Every effort has been made to contact the relevant rights holders of the images contained in this publication. If there have been any accidental omissions, please notify the publisher to correct for any future reprint.

6 NASA

A CONSTANT COMPANION

Gazing at the Moon: Observations and Observatories

13 © Historic England Archive
14 Sonia Halliday Photo Library/Alamy Stock Photo
15 PhotoStock-Israel/Alamy Stock Photo
16 National Maritime Museum (NMM), Caird Fund (BHC1812)
17 Keystone Pictures USA/Alamy Stock Photo

Using the Moon

19 NMM (AST0051.2)
20–21 © Museum of the History of Science, University of Oxford (31528)
20 bottom; NMM, Caird Fund (AST0438)
22 NMM (NVT/40)
23 NMM, Caird Collection (NAV0126)
24 Library of Congress, Geography and Map Division
25 NMM (LOG/M/1)
26 NMM (PBN1982)
27 NMM (ZBA5693)
29 NMM (NAV1803)
31 NMM (ZBA7662)
32 Private collection

'Never Point at the Moon!': Lunar Lore across the Millennia

35 © Phil Hart
36 Age Fotostock/Alamy Stock Photo
37 © The Trustees of the British Museum. All rights reserved.
38 Courtesy of Petrie Museum of Egyptian Archaeology, UCL
39 Metropolitan Museum of Art, New York (1971.272.15)
40 © Victoria and Albert Museum, London
42 Photo © RMN-Grand Palais (musée du Louvre)/Hervé Lewandowski
43 left; Archaeological Museum, Milan, Leemage/Corbis via Getty Images
43 right; National Museum of Antiquities, Leiden
44 © Victoria and Albert Museum, London
46 Science Museum, London
48 Hammer Films/Ronald Grant Archive
49 Metropolitan Museum of Art, New York

The Moon in China

51 Getty Open Content Program
53 Rijksmuseum, Amsterdam

Lunar illumination in the Art of Africa

54, 55 National Museum of African Art, Smithsonian Institution, gift of Walt Disney World Co., a subsidiary of The Walt Disney Company. Object ref. 2005-6-226 (page 54), 2005-6-45 (page 55). Photographs by Franko Khoury
56 left; National Museum of African Art, Smithsonian Institution, acquisition grant from the James Smithson Society. Object ref. 84-6-6.1. Photograph by Franko Khoury
56 right; National Museum of African Art, Smithsonian Institution, gift of Walt Disney World Co., a subsidiary of The Walt Disney Company. Object ref. 2005-6-113. Photograph by Franko Khoury
58 National Museum of African Art, Smithsonian Institution, estate of Barbara and Joseph Goldenberg. Object ref. 2011-4-2. Photograph by Franko Khoury
59 National Museum of African Art, Smithsonian Institution. Object ref. 86-12-14. Photograph by Franko Khoury
60 © El Anatsui, 1993. National Museum of African Art, Smithsonian Institution, purchased with funds provided by the Smithsonian Acquisition Program. Object ref. 96-17-1. Photograph by Franko Khoury
61 © Adebisi Fabunmi, 1960. National Museum of African Art, Smithsonian Institution, bequest of Bernice M. Kelly. Object ref. 98-20-6. Photograph by Franko Khoury
63 © Bruce Onobrakpeya, 1967. National Museum of African Art, Smithsonian Institution, bequest of Bernice M. Kelly. Object ref. 98-20-3. Photograph by Franko Khoury

A Place that Exists Only in Moonlight: The Moon as Muse

64 Photo © John McKenzie. Courtesy of the artist and Ingleby Gallery, Edinburgh
65 © The Samuel Courtauld Trust, The Courtauld Gallery, London
66 © Tate, London, 2019
67 © Walker Art Gallery, National Museums Liverpool
68 © The Fitzwilliam Museum, Cambridge
70 © Darren Almond. Courtesy of White Cube
72–73 Photo © Jonathan Greet. Courtesy of the artist and October Gallery, London and Factum Arte, Madrid

The Moon and Music

75 Pictures Now/Alamy Stock Photo
77 Mary Evans/Grenville Collins Postcard Collection
78 DWD-Media/Alamy Stock Photo
79 Mary Evans/Iberfoto

THROUGH THE LENS

The Telescopic Revolution

82, 83 West Sussex Record Office, Chichester, Petworth House Archives. Reproduced by kind permission of Lord Egremont
84 NMM, Caird Collection (BHC2700)
85 Wellcome Collection
87 NMM, Airy Collection (PBG2055)

Mapping the Moon

89 NMM, Airy Collection (PBG2055)
90 Courtesy of History of Science Collections, University of Oklahoma Libraries
91 NMM, Airy Collection (PBG2055)
92 NMM, Caird Fund (GLB0140)
93 Science Museum/Science & Society Picture Library
94 NMM (PBB2692)
96 Crawford Collection, Royal Observatory Edinburgh
97 NMM, Airy Collection (PBG0909/1)
98 Cambridge University Library (S696.a.93.6)
99 NMM (REC/69/1:81)
100 NMM (ZBA4573.10)
101 NMM (GLB0178)

Portraying the Moon: an Artist's Eye

103 Private collection
104 NMM, Caird Collection (PAJ3148)
105 Science Museum/Science & Society Picture Library

106 Royal Astronomical Society/Science Photo Library
107 Museum of the History of Science, University of Oxford. (11958)

By the Light of the Moon
109, 113 National Science & Media Museum/Science & Society Picture Library
112 Metropolitan Museum of Art, New York
114, 116 Science Museum/Science & Society Picture Library
117 Courtesy of National Gallery of Art, Washington

The Moon through the Eyes of the Apollo Astronauts
119–121 NASA

Revealing the True Colours of the Moon
123–125 Tom Kerss

DESTINATION MOON

The End of the Beginning: Imagining the Lunar Voyage
129 Bibliothèque Nationale de France
131 Houghton Library, Harvard University (STC 11943.5)
132 Houghton Library, Harvard University (FC6 C9938 Eg659s)
133 Courtesy of Göttingen State and University Library
134 Reproduced by kind permission of the Syndics of Cambridge University Library (1877.7.304)
135 Private collection
136 top; © Lobster Films/Fondation Groupama Gan/Fondation Technicolor/2010
136 bottom; Courtesy of BFI National Archive
138 © Victoria and Albert Museum, London
139 George Pal/Ronald Grant Archive

Project Moon: Satirising our Satellite
141 NMM (ZBA4565)
142, 143 Science Museum/Science & Society Picture Library
145 © The Fitzwilliam Museum, Cambridge

Life on the Moon, Newspapers on Earth
147 © The Trustees of the British Museum. All rights reserved
148 top; © Museum of the History of Science, University of Oxford (25152)
148 bottom; Nantucket Historical Association
150 Permission of Llyfrgell Genedlaethol Cymru/The National Library of Wales
151 Science Museum/Science & Society Picture Library

Fly Me to the Moon: From Artistic Moonscape to Destination
153 Photo © National Museums Scotland
154 top; © Victoria and Albert Museum, London
156 NMM (PBB2692)
158 UFA/Ronald Grant Archive
159 bottom; Courtesy of BFI National Archive
161 © Vija Celmins, courtesy of Matthew Marks Gallery. Image: The Museum of Modern Art, New York/Scala, Florence
162 Science Museum/Science & Society Picture Library
164 © 1963 Kiki Kogelnik Foundation. All rights reserved
165 © Robert Rauschenberg Foundation/VAGA at ARS, NY and DACS, London 2019. Collection: National Gallery of Art, Washington
167 © Peter Phillips. Image: © Tate, London, 2019

Fashion and the Moon Landings
169 Courtesy of Bill Ayrey, ILC Dover
170 Interpress Paris/ullstein bild via Getty Images
171 Bettmann/Getty Images
173 © Victoria and Albert Museum, London

Space Cadets: Toys and Material Culture from Sputnik 1 to Apollo 11
175 © Victoria and Albert Museum, London
176 Mattel/Solent News/REX/Shutterstock
177 © Victoria and Albert Museum, London

FOR ALL MANKIND

The Scientists and the Moon
181 top; Courtesy of Lunar & Planetary Institute, Houston, Texas
181 bottom; NASA
182 NASA/GSFC/Arizona State University
184 USGS Astrogeology
185 top; USGS Astrogeology
185 centre; NASA/GSFC/Arizona State University
185 bottom; USGS Astrogeology
188 top; NASA
188 bottom; www.virtualmicroscope.org
189 NASA/MSFC/Renee Weber
190–191 NASA
193 NASA/GSFC

So What If It's Just Green Cheese? The Moon on Screen
195 Courtesy of BFI National Archive/MGM
196 top; Courtesy of BFI National Archive/Eon Productions
196 bottom; Courtesy of Ronald Grant Archive/Eon Productions
199 Liberty Films/Ronald Grant Archive

Poetry and the Moon: Moon Vehicles, Moon Metaphors
203 NASA
204 Penguin Books, 1976
205 RAI/Ronald Grant Archive

Whose Moon? The Artists and the Moon
207 Courtesy of the artist
208 © Cristina de Middel/Magnum Photos
210 © Tom Hammick. Courtesy of Flowers Gallery London and New York. Bridgeman Images 2019, all rights reserved
211 © Chris Ofili. Courtesy of Victoria Miro Gallery/Victoria and Albert Museum, London
212–213 Courtesy of Leonid Tishkov and Boris Bendikov

The Geopolitics of the Moon
215–217 NASA
219 © 2019. The Clangers/Smallfilms/Peter Firmin
221, 222 NASA
224 Granger Historical Picture Archive/Alamy Stock Photo

EPILOGUE

231, 232, 234, 235 NASA

ENDNOTES

243 Dior

ENDPAPERS

NMM (ZBA4573.10)

Acknowledgements

As editors of this book, we have accrued many debts of gratitude. Our first thanks must go to the contributors for their outstanding chapters, each of which speaks to the broader themes of the publication and the exhibition it accompanies. It will be for the reader to decide whether this book is greater than the sum of its parts, but there is no doubting that the parts themselves are of exceptional quality.

The Moon sits alongside a major temporary exhibition of the same name, and as curators of the exhibition as well as editors of the book, we would like to thank the formidable team who worked with us on the project: Sandra Adler, Gareth Bellis, Emily Churchill, Celine Dalcher, Lisa Evans, Nicola Fleming, Nadine Fleischer, Myrthe Huijts, Sarah Kavanagh, Emmanuelle Largeteau, Helena Liszka, Robin Marsden, Kristian Martin, Claire Mead, Goizane Mendia-Rios, Ben Raithby, Anna Rolls, David Rooney, Bethia Varik and Marieke de Veer. It was Sarah Wood's unerring inquisitiveness, generosity and humour that brought everything together.

The exhibition's lenders whose objects are also represented in this volume must be thanked: The British Film Institute; the British Library; Lord Egremont; the Fitzwilliam Museum, Cambridge; Foster and Partners; the Harryhausen Foundation; the Museum of the History of Science, Oxford; NASA; National Museums Scotland; the Royal Astronomical Society; the Royal Observatory Edinburgh; the Science Museum Group; Senate House Library; the National Air and Space Museum, Smithsonian Institution; Tate; the Victoria and Albert Museum; Wellcome Collection, and several private lenders.

In addition, we are grateful to artists Darren Almond, El Anatsui, Cristina de Middel, Vincent Fournier, Tom Hammick, Aleksandra Mir, Katie Paterson, Peter Phillips, Chris Ofili, Larissa Sansour, Joe Tilson, Leonid Tishkov, and their galleries: Ingleby Gallery, Parafin, October Gallery, White Cube and Victoria Miro.

We would also like to thank Silke Ackermann, William 'Bill' Anders, Geoff Belknapp, David Blayney Brown, Lucy Blaxland, Alison Boyle, Beata Bradford, Dallas Campbell, Qin Cao, Lucy Carson, Ian Crawford, Cassie Davies-Strodder, Karen Demuth, David DeVorkin, Ellen Grace, Adam Greenhalgh, Alfred Haft, Michael Light, Elenor Ling, Henry Little, Alison McCann, Valerie Neal, Michael Neufeld, David Packer, Sian Prosser, Andrew Smith, Suzanne Smith, Joy Sleeman, Ernst Vegelin van Claerbergen, Sophie Waring and Adrian Whicher. Melanie Vandenbrouck would like to extend her special thanks to Richard Willis and Raphael for their patience.

The quality of the images contained in the book is testament to the expertise of the Museum's Conservation department, the skilled work of its Photo Studio, and the thoroughness and perseverance of Andrew Tullis in sourcing images from multiple institutions. At HarperCollins, Sheena Shanks was a dedicated and efficient editor. Most of all, we would like to thank Taragh Godfrey, whose prodigious organisation, mastery of detail and unstinting patience were invaluable as she steered the book through to completion.

Schuckburgh
Atlas
WILLIAMS
HMW
MAKE
De